Fun with Maths

PETER PATILLA

KINGfISHER

KINGFISHER
Kingfisher Publications Plc
New Penderel House
283-288 High Holborn
London WC1V 7HZ

First published in hardback
by Kingfisher Publications Plc in 1998
ISBN 0 7534 0301 3
2 4 6 8 10 9 7 5 3 1

First published in paperback
by Kingfisher Publications Plc in 1998
ISBN 0 7534 0281 5
2 4 6 8 10 9 7 5 3 1

Material in this edition was previously published in 1997
by Kingfisher in four separate volumes: *Guide to Good Maths: Adding, Substracting, Multiplying* and *Dividing*

A CIP CATALOGUE RECORD FOR THIS BOOK
IS AVAILABLE FROM THE BRITISH LIBRARY

Editor: John Grisewood
Illustrations: Terry McKenna
Design: Robert Wheeler Design Associates
Phototypeset by Southern Positives and Negatives (SPAN), Lingfield, Surrey
Printed in Hong Kong/China

CONTENTS

ADDING

Adding

Adding with small numbers

Magic squares

It is thought that magic squares were first discovered about 2200 BC in ancient China by Emperor Yu. They were called Loh-Shu and dot patterns were used instead of numbers.

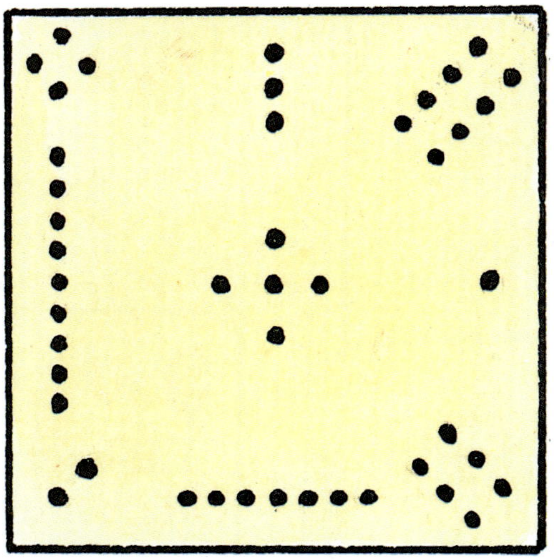

A Loh-Shu
It is 'magic' because the number of dots in each line, column and diagonal is the same.

LOH-SHU

	3	
		1
	7	

Can you complete these magic squares? Each row, column and diagonal must total 15.

	1	
		2

Totalling scores

Adding the numbers you score in games is an important skill. Can you find this score by adding the numbers in your head?

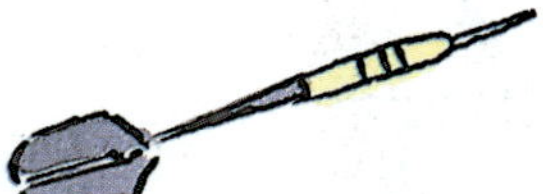

Addition walls

Look closely at this addition wall.
Can you see how the number
on the top brick is found?

Copy and complete these addition walls.
One of the walls has lots of possible answers.

Adding digits investigation

Here are all the ten possible digits:

You can use some of these digits to investigate number problems.

Choose these four
digit cards.

Make totals from the cards.
You can only use + and =.
Here are some totals.

4+5=9   4+2+7=13   5+2+7+4=18

Can you find some more?

Choose four digit cards yourself.
Investigate the totals you can make.

Adding games

Patience game: Make Twenty

Use a pack of playing cards with all the picture cards taken out.

Shuffle the pack and put it face down on the table.
Turn over one card at a time and place it face up in line.

When you see a set of 'consecutive' cards which totals 20 gather the set up and close any gaps.
A set can have as many cards in it as you want but the total of the set must be 20.
Try to end up with as few cards in the line as you can.

Twenty-one game

A game for two players.
Use a set of digit cards 1–9.

Toss to see who goes first.
Take turns to choose a digit card and place it in the centre of the table. The total of this set of cards in the centre is important.
The aim of the game is either to make the total exactly 21 or to make your opponent go over 21.
If you total 21 score one point.
If you go over 21 score zero points.
Play several rounds to decide the winner.

Cover-up game

There is a popular dice game which uses simple adding although some skill is needed when the game is being played.

Several players can take part.
The first player rolls two dice and finds the total.
One or more numbers are then covered which match the total.

For example  8 may be used to cover 8 or 6 2 1 2 5

Only one dice is rolled when the remaining numbers total less than 6.
The turn ends when no more numbers can be covered up by the dice total.

The winner is the one who covers up all the numbers or who has the lowest total uncovered at the end.

Joining dots game

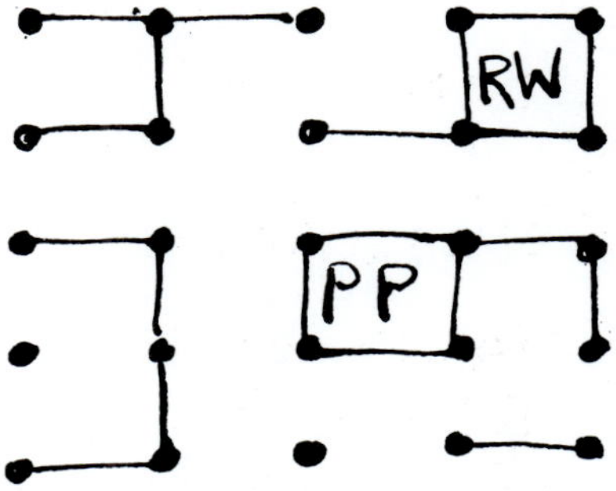

Lots and lots of people play joining the dots game to pass the time away.
Two dots are joined by a straight line.
Whoever can complete a square writes their initial inside the square.
The player who completes most squares is the winner.

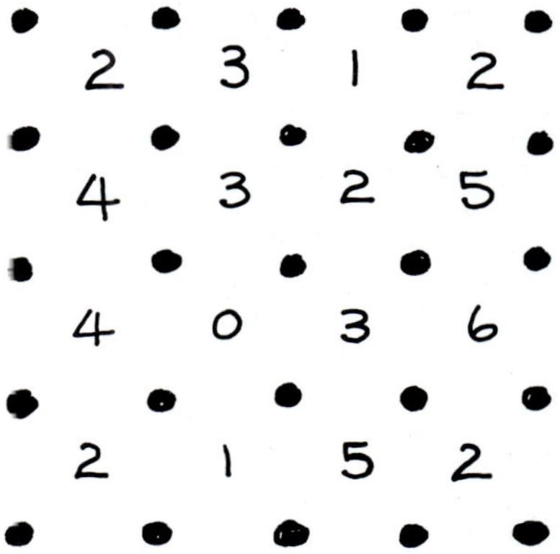

More skill can be added to this game by putting numbers inside the dots.
The game is played as usual with each player trying to complete as many squares as possible.
When all the squares are complete each player totals the numbers in their squares.
The winner is the player with the highest total.

Investigating addition

When trying some of these investigations remember that there are lots of different answers to some of the problems. It is important to set your work out neatly so that other people can see what you have been doing.

Consecutive sums

Here are some sets of consecutive numbers and their sums.

1 + 2 + 3 + 4 SUM 10

11 + 12 + 13 SUM 36

8 + 9 SUM 17

Try and find sums, up to 50, which you can make by adding two consecutive numbers.

Which sums can you make by adding three consecutive numbers?

Are there any sums you cannot make by adding consecutive numbers?

Square number totals

Use a set of number cards 1–10.
Put the cards into pairs.
Each pair must add up to
a 'square number'.

Here are some square numbers:

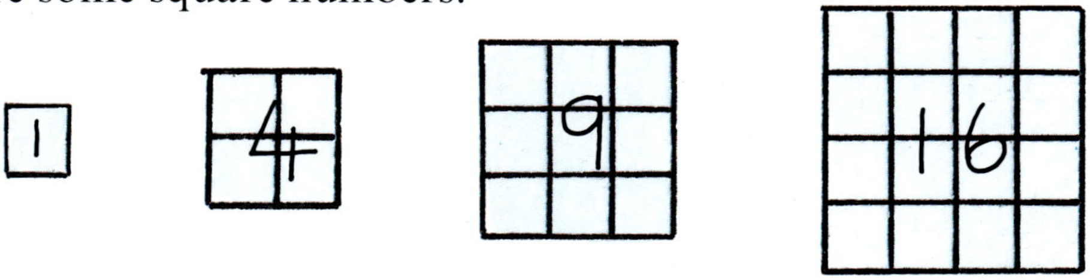

How many pairs of cards can you find?

Try making sets of three cards which add up to square numbers.

Adding light bars

Use a calculator.
Add up the light bars which make the numbers.

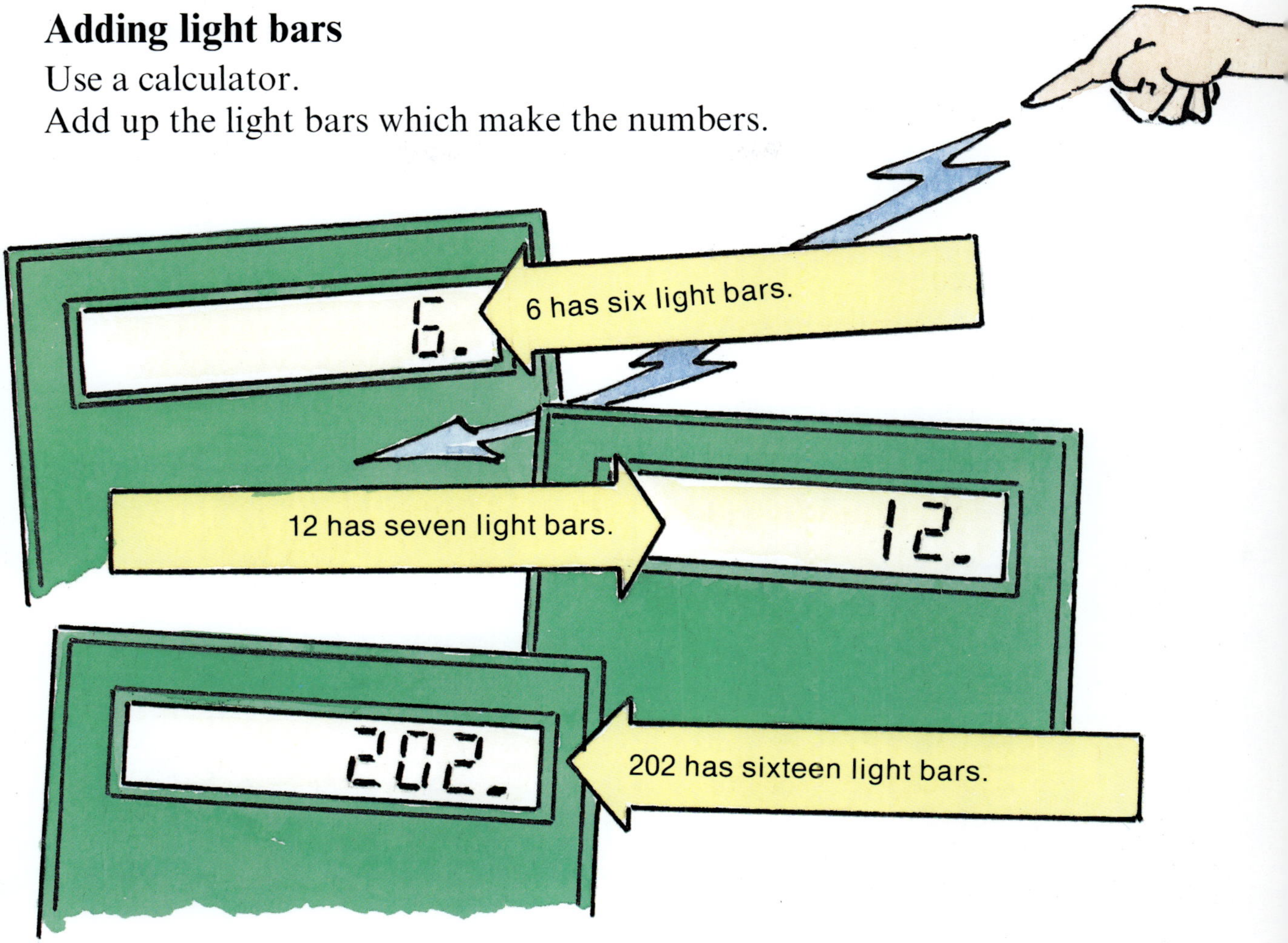

How many different numbers can you make which have eight light bars?

Explore different numbers of light bars.

Dice totals

Use one dice.
Roll the dice several times.
Investigate which number is most likely to turn up.

Use two dice.
Roll them and find the total.
Investigate which total is most likely to turn up.

Addition puzzles

Your addition skills can help you solve puzzles.

Alphabet codes

Find the answers to the sums.
Use the answer code to find the letters.
Then, rearrange the letters to find the name of a country.

9+8	12+7	14+8	15+12	13+7	14+12	16+8

A 19	B 18	C 25	D 15	E 20	F 12	G 24	H 21	I 11	J 31	K 16	L 29	M 22
N 26	O 35	P 34	Q 23	R 27	S 14	T 13	U 32	V 10	W 33	X 28	Y 17	Z 30

What is the name of the country?
Make up some alphabet codes of your own.

Word sum puzzles

In this sum each letter stands for a number.
The letter D = 5.
Can you find what the other letters stand for?

```
  CAN
+ ADD
-----
 GOOD
```

Now try this
Word Sum Puzzle.

The letter O = 9

```
  LOST
+   IT
------
  LOOK
```

Addition square puzzles

Which numbers go in the empty squares?
The horizontal sums must be correct.
The vertical sums must be correct.

Word total puzzles

Each letter of the alphabet can have a value.

A	B	C	D	E	F	G	H	I	J	K	L	M	N	O	P	Q	R	S	T	U	V	W	X	Y	Z
1	2	3	4	5	6	7	8	9	10	11	12	13	14	15	16	17	18	19	20	21	22	23	24	25	26

ANT is worth 1 + 14 + 20 = 35

ZEBRA is worth 26 + 5 + 2 + 18 + 1 = 52

Can you find an animal worth less than ANT?
Can you find an animal worth more than ZEBRA?

Adding larger numbers

How good are you at adding 2-digit numbers?

Euler's square

In the 18th century a Swiss mathematician called Leonard Euler created this special square.

All the numbers from 1 to 64 appear in the square.

What is special about the totals in:

each row?
each column?
each of the four squares?

Explore other totals in Euler's special square.

How good are you at adding 3-digit numbers?

Consecutive numbers

Pairs of consecutive 3-digit numbers have been added together to make these totals.

Can you find the pairs of numbers?

Dice game

Roll a dice

Each player writes the number rolled in any of the boxes (do not let the other players see where you put the numbers!)
Roll the dice again and put the number in another box.
Keep doing this until all the six boxes are full.
Each player finds the total of their sum.
The person with the total nearest to 500 wins.
Play several games.

All digit adds

Each of these sums uses all the digits 1–9.

$$\begin{array}{r} 654 \\ +\ 318 \\ \hline 972 \\ \hline \end{array} \qquad \begin{array}{r} 586 \\ +\ 341 \\ \hline 927 \\ \hline \end{array} \qquad \begin{array}{r} 216 \\ +\ 738 \\ \hline 954 \\ \hline \end{array}$$

Can you find some sums of your own which use all the digits 1–9?
This is quite a challenge!

Adding with grids

Addition grids

Here is a simple addition grid.
Can you see how it works?

+	3	6	8
2	5	8	10
5	8	11	13
7	10	13	15

+	5		9	
2	7		11	13
	9			
			18	
10	15	17		
			21	

Here is an addition grid with some numbers missing. Copy it and find the missing numbers.

Problem grids

Some grids make you think!

The total of each row is shown.
The total of each column is shown.
What could the numbers in the grid be?

There may be more than one answer.

Triangle grids

Grids do not have to be made from squares.

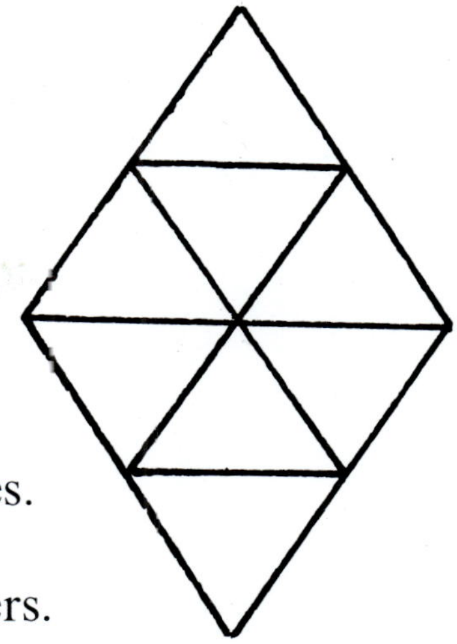

Here is a grid made from triangles.
The numbers 1–8 are missing from the grid.
When a pair of numbers touch their total must be ODD.
Numbers do not touch at corners only on sides.
Copy the grid and try to find where to put the numbers. There are several possible answers.

Digit grids

Grids which have lots of digits in them can be used for making up problems. Here is a chain of digits which add up to 20.

4	8	7	2	9	1	3	1
1	2	4	9	5	3	5	6
7	2	3	1	5	4	3	5
1	8	7	6	1	1	5	6
4	3	1	2	6	2	4	3
1	2	9	3	4	1	2	1
3	3	4	8	5	3	6	4
2	5	2	7	9	4	8	2

Can you find any more 20 chains?
What is the longest 20 chain you can find?

Choose a different total and explore chains which make your total.

Addition problems

Addition problems come in all shapes and sizes. Some problems are simple and straightforward, others are there to make you think. See how you get on with these problems.

Dice problem

Look carefully at these pictures of dice. Some of the dice spots are touching the table.
Can you find out how many spots touch the table in each set of dice?
It will help if you examine how a dice is numbered.

Word problem

I have 148 stamps from America, 276 from Spain and 236 from France. How many stamps do I have altogether?

Mental problems

What do two quarters add up to?
What is half add a quarter?
What do two halves make?

Digit problems

Use the digits 1–9.
The middle number is odd.
Each pair of numbers has an odd total.
The five vertical numbers have an odd total.
The five horizontal numbers have an odd total.
PROBLEM: Where do the numbers 1–9 go?

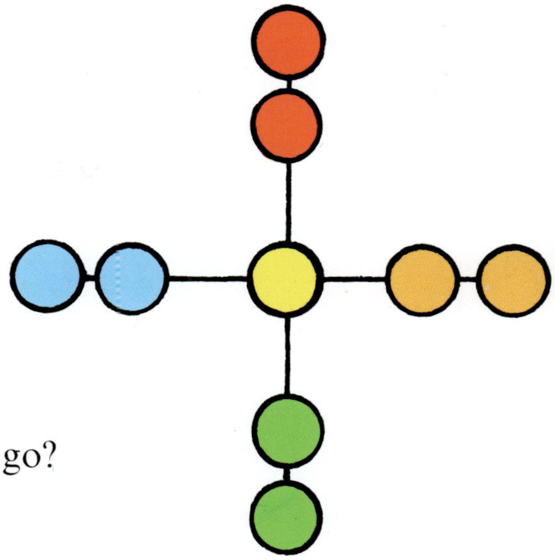

Calculator problems

How many days has it been since your fifth birthday?

How many days have passed since you were born?

How many drinks do you think you have in one year?

Real problems

Calculator Additions

A calculator can be used to help you add up large numbers. It can also be used to explore and experiment with addition of numbers.

Limited touches

The keys must be touched SIX times, no more, no less.

Here are two ways an answer of 20 can be made with SIX touches.

8 + 9 + 3 =

1 0 + 1 0 =

Can you find some more ways of making 20 with SIX touches?
Now try and make 20 with FIVE touches.

Target

3 4 7 + =

You can only touch these keys

Each key can be touched as often as you like.
They can be touched in any order.
Not all of them have to be touched.

13 can be made like this 3 + 3 + 7 =

17 can be made like this 7 + 4 + 3 + 3 =

Try to make as many numbers between 10 and 30 as you can.

Forbidden key

You are NOT allowed to touch the [8] key.
Here are two ways of finding the answer to 18 + 6 without touching the [8] key.

17 + 1 + 6 =   9 + 9 + 6 =

Try to find different ways of doing these sums without touching the [8] key.

48 + 3   28 + 8   84 + 16   88 + 8

ZAP the digit

A game for two players.

Put a large number on the display.

To ZAP a digit means to replace it with a zero by doing an addition sum. Each player takes turns to choose a digit to ZAP.

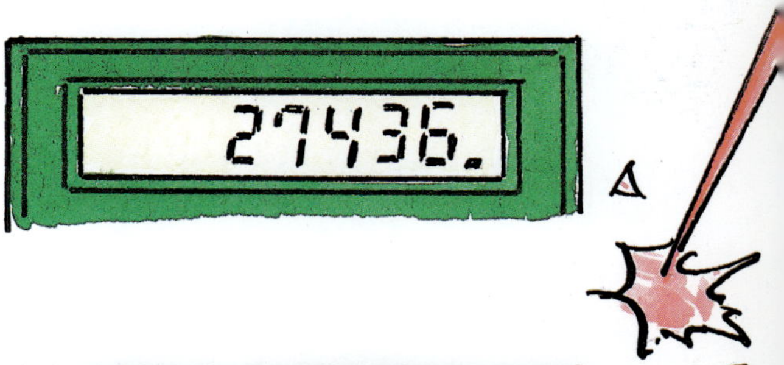

You can ZAP 4 by adding 600.
The 4 becomes zero.

You can ZAP 3 by adding 70.
The 3 becomes zero.

Who will make the last ZAP?

Missing numbers in additions

Some algebra work is concerned with finding 'missing numbers'. The missing numbers can be in EQUATIONS, in SUMS, in PROBLEMS. Try to find the missing numbers in the following activities.

Simple equations

Each of these equations has a missing number.
There is only one possible answer to each equation.

$\square + 9 = 17$ $15 + \square = 24$ $13 + 12 = \square$

Open equations

Sometimes equations can have several possible answers.

There are two missing numbers in each of these equations.
How many pairs of numbers can you find for each equation to make it true?

$\square + \triangle = 10$ $\square + 3 = \triangle$ $7 + \square + \triangle = 20$

Addition sums

Each star stands for a missing digit.
Use your addition skills to work out which digit each star stands for.

$$\begin{array}{r} 43 \\ +\ 2* \\ \hline *1 \end{array} \qquad \begin{array}{r} *6 \\ +\ 98 \\ \hline *5* \end{array} \qquad \begin{array}{r} 38 \\ +\ ** \\ \hline 66 \end{array}$$

Addition grids

+		
	7	8
	10	11

Some numbers are missing from the outside of this addition grid.
What could the numbers be?
Is there only one possible answer?

Word problems

Addition sequences

Find the missing numbers in these addition sequences.

45 50 ** 60 ** 70 ** 80

34 ** 42 46 ** ** 58 **

** ** 44 51 58 ** ** **

Missing fractions

Find the missing fractions.

$\frac{1}{2} + \square = \frac{3}{4}$ $\frac{1}{4} + \square = 1$ $\frac{3}{4} + \square = 1\frac{1}{2}$

Number Patterns in Additions

Investigating and exploring number patterns is an important part of algebra.
Have fun with these different types of number patterns.

Triangle patterns

Find the total of each TRIANGLE pattern. What do you notice?

1	2	3	4	5	6	7	8	9
1	2	3	4	5	6	7	8	
1	2	3	4	5	6	7		
1	2	3	4	5	6			
1	2	3	4	5				
1	2	3	4					
1	2	3						
1	2							
+1								

	9	8	7	6	5	4	3	2	1
		8	7	6	5	4	3	2	1
			7	6	5	4	3	2	1
				6	5	4	3	2	1
					5	4	3	2	1
						4	3	2	1
							3	2	1
								2	1
+									1

Palindromes

Palindromes are numbers which read the same forwards and backwards.

44 121 7667 14541 492294

Which is the largest palindrome under 1000?

Write a 3-digit number.	329
Reverse it and add.	+ 923
	1252
Repeat this until you get	+ 2521
a palindrome answer.	3773

Try with different 3-digit numbers.
Do you always get a palindrome?

Odds and evens

Which kind of number do you get if you add:

Two ODD numbers?

Two EVEN numbers?

An ODD number to an EVEN number?

An EVEN number to an ODD number?

Three EVEN numbers?

Three ODD numbers?

Mental additions

Numbers can be broken up into smaller units which give the same total:

$$
\begin{aligned}
46 &= 40 + 6 \\
&= 30 + 16 \\
&= 20 + 26 \\
&\text{etc}
\end{aligned}
\qquad
\begin{aligned}
145 &= 100 + 40 + 5 \\
&= 100 + 20 + 20 + 5 \\
&= 50 + 50 + 40 + 5 \\
&\text{etc}
\end{aligned}
$$

We can use this fact to help us add up in our heads.

$$
\begin{aligned}
56 + 32 &= 50 + 6 \;+\; 30 + 2 \\
&= 80 + 8 \\
&= 88
\end{aligned}
$$

Try to add these up in your head:

Adding with money

Adding skills are needed when handling money.
Use your adding skills to find the answers to these tricky money problems. The money we use is the Drat and Petro. There are 100 petros (p) to a Drat (D)

Coin totals

Which totals up to 20p can be made using these coins.
A coin may not be used more than once.

Five coin totals

Add coins to make totals up to D1.
Which totals need more than FIVE coins to make them?

Coin game

A game for two players.
Use 10p and 20p coins.

Decide who will go first. One uses heads, one uses tails.
Take turns to place a 10p or a 20p coin on the grid.
The first player to make a line which totals 50p wins.
The line can be horizontal, vertical or diagonal.

Adding with measures

Adding skills need to be used when we are measuring length, weight, capacity, and time.

How heavy?

The weights used to find the mass (weight) of each parcel are shown. How heavy is each parcel?

Weigh some objects of your own.

Finding perimeters

Perimeter is the distance measured all the way round a shape.
Find the perimeters of these shapes:

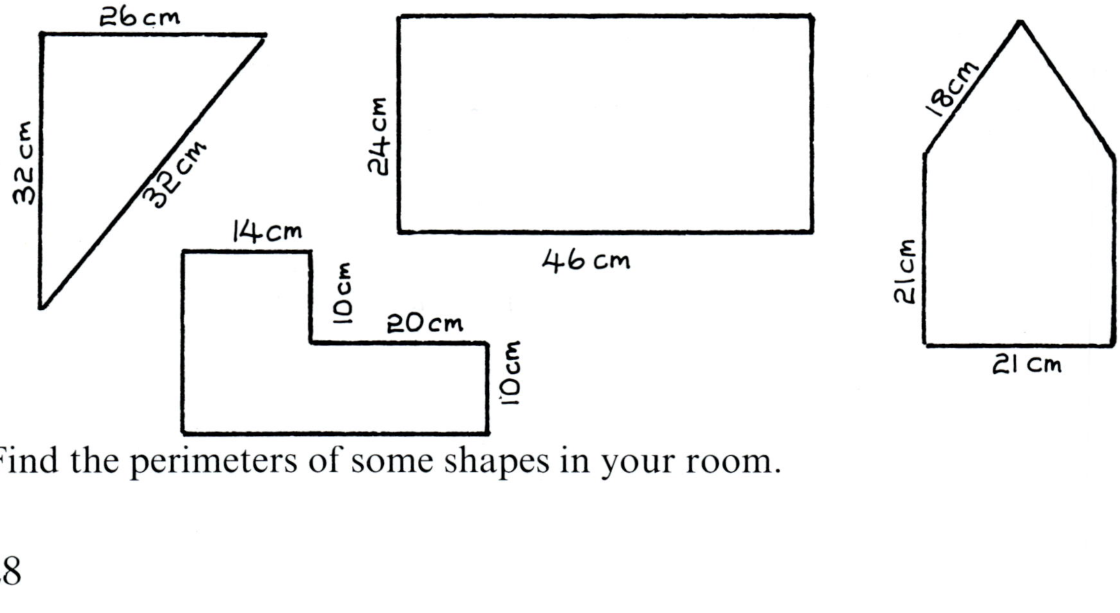

Find the perimeters of some shapes in your room.

Extra time

Look at each watch. What will the time be 45 minutes later?

Metric totals

Find the totals of these metric measures.

1.45m, 2.75m

1.850Kg, 2.560Kg

3.700 litres, 2.125 litres

6.5Km, 5.9Km

45mm, 69mm, 38mm

69ml, 36ml, 78ml

Imperial totals

Find the totals of these imperial measures.

11ins, 9ins, 6ins

12oz, 14oz, 13oz

6pts, 7pts, 5pts

12lb, 13lb, 9lb

6ft, 17ft, 21ft

9gal, 23gal, 12gal

Adding and data handling

Information is presented to us in all sorts of ways:

Graphs Lists Diagrams

Tables Charts Pictures

Use your skills to interpret the graphs and find the answers.

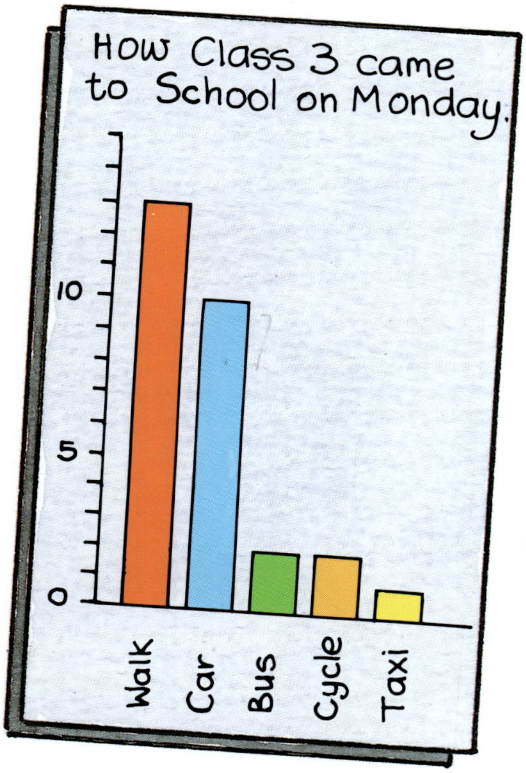

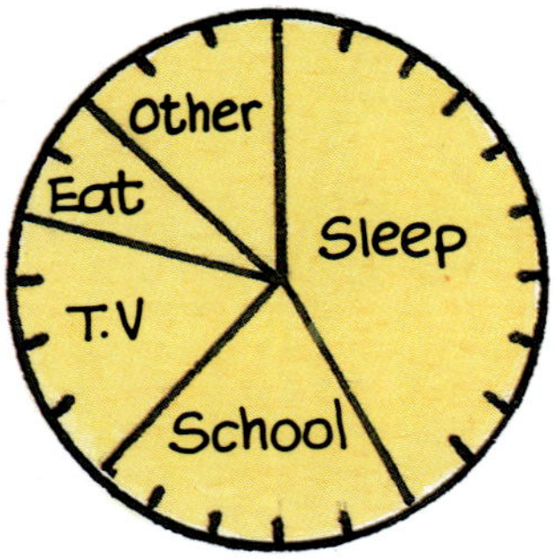

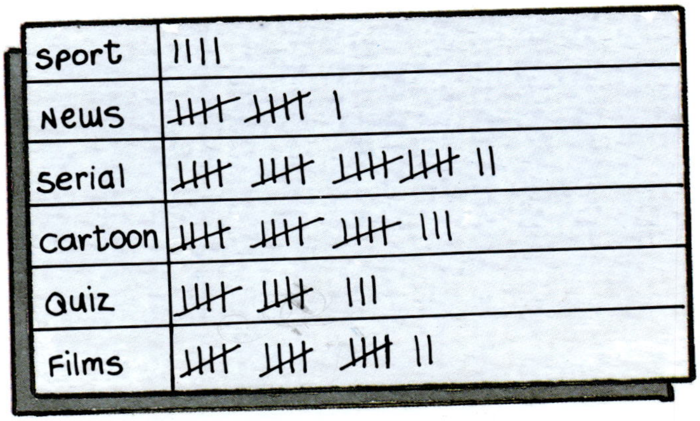

How many children are in class 3?

What is the average number of hours class 3 were awake?

How many children arrive at school under their own steam?

How many children watched films on Monday?

On average how many hours of TV were watched?

SUBTRACTING

Subtracting

Taking away

There are many different kinds of subtraction. One sort of subtraction is called 'taking away', when what you started with becomes smaller.

Try these take away activities.

Abacus numbers

Here are some numbers on an abacus.

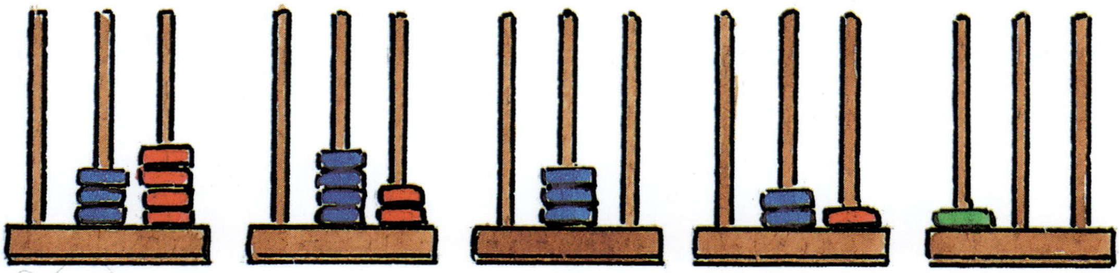

Take away 12 from each abacus number and draw the answers on your own abaci.

Halving

Take away half of each HTU (hundred, ten, unit) number and draw what is left.

Nim

Nim is a game which involves taking away. It needs two players and 20 small objects to start with.

Take turns to remove 1, 2, 3 or 4 of the objects.
The player who takes the last object or objects wins.

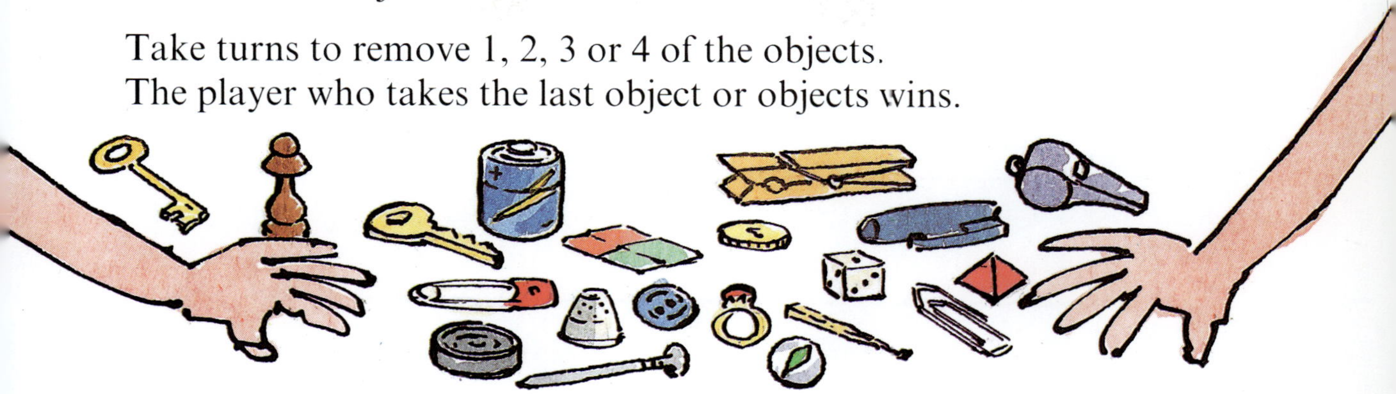

Play several games.
Can you find a way of winning?

Try changing the number of objects you start with.

Smaller measures

Sometimes we take away from lengths and quantities.

Cut off 45cm.
What is left?

Drink 500ml.
What is left?

Eat 125g.
What is left?

Spend D1.45
What is left?

Difference

There is a type of subtraction called ‘difference’. Nothing is actually taken away when finding the difference between two numbers. It is comparing two numbers finding how many more, or less, one number is than the other.

> The difference between 6 and 10 is 4.
> The difference between 5m and 2m is 3m.

Try these difference activities.

Pairs

Find the difference between each pair of coins.

Card game

Use a pack of cards without the picture cards.
Three cards are dealt to each player.
The rest of the cards are placed face down.
Players take a card from the pack in turn.
When a difference of 3 can be made that pair of cards is placed face down.
Another turn can be had whenever a pair is made.
The winner is the player who gets rid of all their cards first or who has the smallest total in their hand at the end.
Play the game again but change the difference.

Digit cards

Make digit cards 1 to 9.
Find a home for each card.
The stars show the difference between adjacent numbers.

Difference challenge

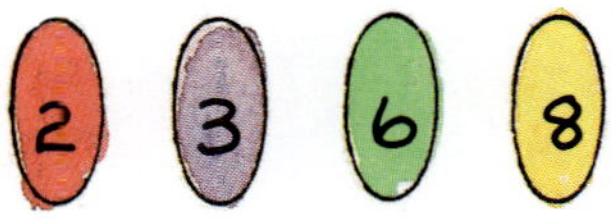

You are only allowed to use these four digits.
How many different answers can be made by finding differences?

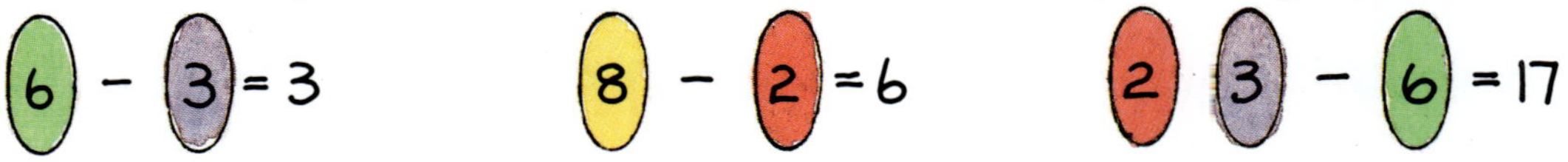

Domino difference

These domino totals have a difference of 4.

Can you draw other domino pairs which have a difference of 4?

Complementary addition

Another type of subtraction is called 'complementary addition'. This is often the method used by shopkeepers when counting out your change. Complementary addition is adding on from one number to another.

All change

In the country of Dratonia the currency is as follows:
100 petros = 1 Drat.
There are denominations of D1, 50p, 20p, 10p, 5p, and 1p.
The shopkeeper uses the fewest number of coins she can when giving change. Which coins will she give as change when 5 Drat is offered as payment?

Number chains

Two numbers can be subtracted using complementary addition and number chains.

94 take away 37.

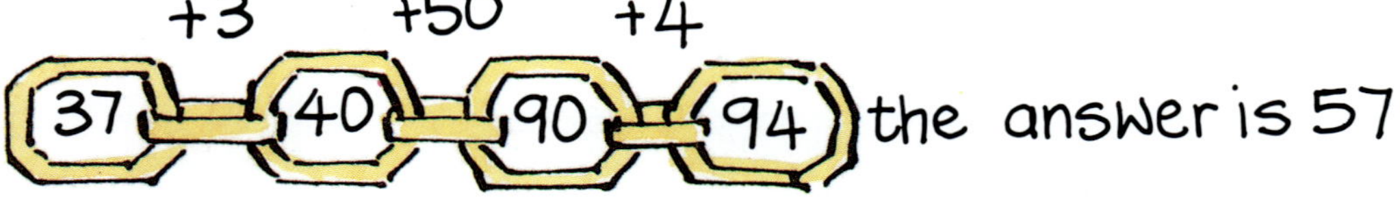

Try these chains yourself.
Different chains are possible between each pair of numbers.

Equations

Use complementary addition skills to find the missing numbers in these equations.

$7 + \square = 15$ $3 + \square = 14$ $5 + \square = 18$

$26 + \square = 32$ $14 + \square = 37$ $28 + \square = 50$

Time check

How many minutes are there between the times shown on each pair of watches?

Round up

What must be added to each of these numbers to round them up to the next 100?

87 132 286 454 638

Subtraction games

Four in a line

A game for two players.
A calculator can be used.
Players take turns to choose two star numbers and find the difference between them.
If the difference is on the grid cover it with a small counter (pieces of card will do).
The first player to get four counters in a straight line, horizontally, vertically or diagonally wins.

19	28	41	10	31	11
32	61	50	63	25	55
47	53	22	44	40	34
62	15	66	23	38	21
60	22	33	14	29	17
54	10	12	74	30	16

Calculator game

Start with 100 on the display.
Take turns to subtract any number which is less than 10.
Who will be the player to reach zero?

Dice game

Players draw empty boxes like this.

Roll a dice.
Each player writes the number rolled in any one of the boxes.
(Do not let the other players see where you put the numbers.)
Roll the dice again and put the number in another box.
Keep doing this until all six boxes are full.
Each player subtracts their two numbers.
The player with the answer nearest to 150 wins.

Nine lives game

Each player begins with nine lives.

Copy the grid.
Join two dots with a straight line.
Try to complete squares.
If you complete a square which has −1 inside it then you lose one of your lives.
The winner is the player with most lives left at the end of the game.

Subtracting large numbers

How good are you at subtracting two and three digit numbers? Try your skills out on these activities.

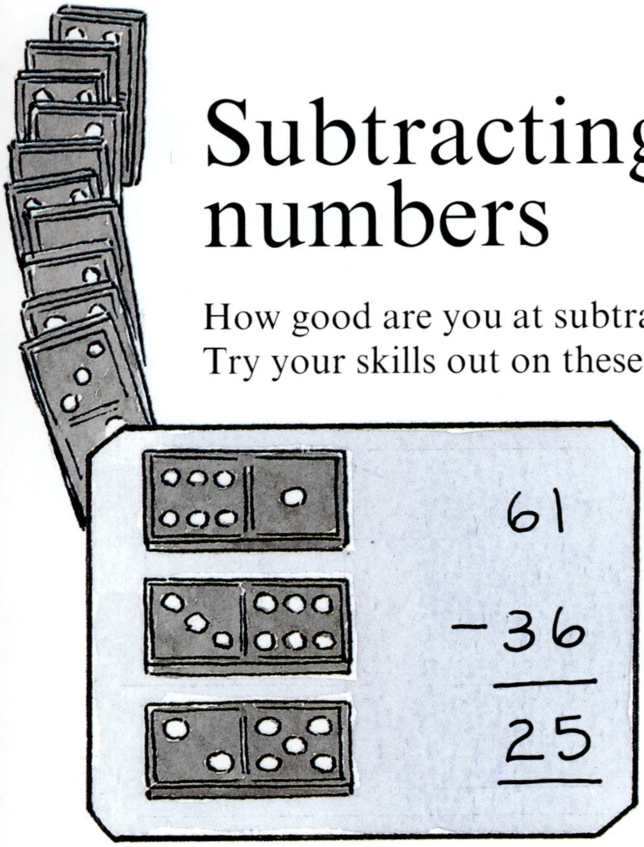

Domino problem

Here is a subtraction problem made from three dominoes. Make some different domino subtractions and draw the answers.

Digit switch

You can use a calculator to help you explore this number pattern activity.

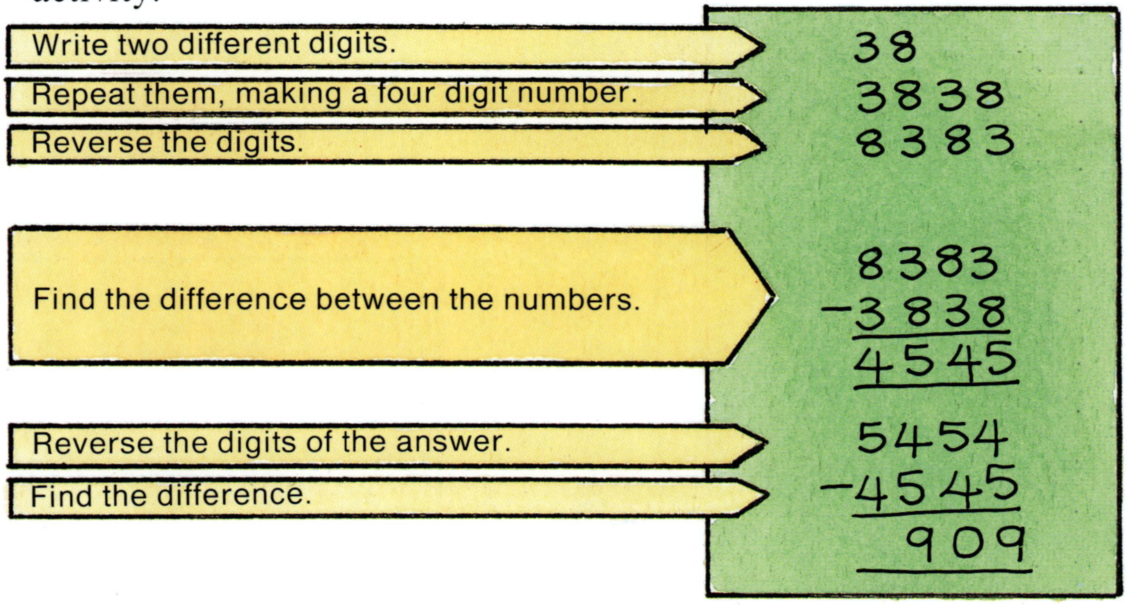

Keep doing this until you have an answer with only three digits.

Repeat with a different pair of starting digits.
Can you guess what the answers will be each time?

Darts scores

Subtraction skills are needed when playing darts
Find each darts total.
Subtract the total from the player's score.

Consecutives

Numbers which come after each other are called 'consecutive'.

□ 23 and △ 24 are consecutive. □ 45 and △ 46 are also consecutive.

Can you find the missing consecutive numbers in these equations?

157 − □ − △ = 100   213 − □ − △ = 100

171 − □ − △ = 100   187 − □ − △ = 100

239 − □ − △ = 100

Subtraction puzzles

Try and solve these puzzles using all your skills of subtraction.

Word teaser

In this problem each letter stands for a number.
The letter E = 4 and S = 6
Can you find what the other letters stand for?

Now try this problem.
The letter H = 9 and E = 4

(There may be more than one answer)

Code wheel

Find the answers to the subtractions.
Use the wheel code to find the letters.
Rearrange the letters to discover the names of some cities.

Subtract 9 from each number
22 26 29 23 18
(*A city in Australia*)

Subtract 7 from each number
21 15 25 17 20
(*A city in the Netherlands*)

Subtract 8 from each number
27 17 25 16 23
(*A city in France*)

Bubble puzzle

Take:
A number from the first bubble.
A sign from the second bubble.
A number from the third bubble.
The correct answer from
the fourth bubble.
Do this till the bubbles are empty.

Mystery names

Look at the subtraction problems.
Write the answers in words.
Line up the words as shown.
Can you discover the mystery names?

34–24	* _ _
25–21	_ * _ _
41–32	_ _ * _
35–27	_ * _ _ _

33–26	* _ _ _ _
37–29	_ _ _ _ *
31–28	_ _ _ * _
32–25	_ _ * _ _
27–18	_ _ _ *
38–29	_ _ * _

Investigating subtractions

When trying some of these investigations remember that there may be lots of different answers to the problems. Don't stop once you have found one possible answer. Do explore and find other different solutions.

Number cards

Make some small cards numbered 1–9.

Use the cards to make a set of subtraction problems.

Here is a set of two problems which uses six of the cards.

9 - 8 = 1   7 - 2 = 5

The cards 3 4 6 have not been used.

How many different sets of subtraction problems can you make using the cards?

Can you make a set of three problems which uses all the cards?

Making answers

Use these digits.

3 5 6 8

Put them into pairs with a subtraction sign between them.

68 - 35   65 - 38

85 - 63

Can you make answers of:

27   33   22

How many different answers can you make using other pairs?

Forbidden keys

The challenge is to find ways of doing these subtraction problems on a calculator without touching the 2 key.

$$\begin{array}{r} 72 \\ -\ 38 \\ \hline \end{array} \qquad \begin{array}{r} 91 \\ -\ 26 \\ \hline \end{array}$$

$$\begin{array}{r} 120 \\ -\ 32 \\ \hline \end{array} \qquad \begin{array}{r} 242 \\ -\ 29 \\ \hline \end{array}$$

Start and finish

Start with the number 20.
Finish with the number 3.
You can only subtract.
How many ways can you find?

20 - 17 = 3

20 - 14 - 3 = 3

20 - 15 - 2 = 3

Calculator subtractions

A calculator is a useful tool in helping you find answers to subtraction problems. It can also be used to help you explore and discover new things about numbers.

Lift off

You can only touch these keys [2] [5] [6] [−] [=]

Each key can be touched as often as you wish.
They can be touched in any order.
Not all of them have to be touched.

Here is one way of making 10. [2] [5] [−] [5] [−] [5] [−] [5]

Make each of these answers for lift off. . .

Time warp

Try going backwards in time.

What was happening. . .

Super change

Change 278 to 19
Touch the fewest number of keys possible.
How many keys did you touch?

Try these super changes.

764 to 28 301 to 85 724 to 56

Subtraction words

Find the answer to 9034 – 1929.

Turn your calculator upside down and you will see something that seeds grow in.

Make up some subtraction problems which will show these words when the calculator is turned upside down.

BIBLE	SLOG	SOLE
hILL	LOOSE	OBOE

Find some calculator words of your own and write subtraction problems which make them.

Subtraction problems

There are all sorts of different subtraction problems. Some are simple and straightforward while others are tricky and make you think.
Try these different sorts of problems.

Bracket problem

When brackets are in a problem you work out the bracket part first.

$17 - (9 - 5) =$

Work out $9 - 5$ first.
The answer to the problem is 13.

$(17 - 9) - 5 =$

Work out $17 - 9$ first.
The answer to the problem is 3.

In these problems the brackets have been missed out.
Can you find where to put them to make each answer 13?

18 − 3 − 2	15 − 4 − 2	24 − 3 + 8
9 + 11 − 7	18 − 8 − 5 − 2	16 − 2 − 9 − 8

Mental Problems

Take away a quarter from one. Subtract half from three quarters. Find the difference between a quarter and three quarters.

Word problems

If 100 petros equals 1 Drat, what change will be received if I spend 345 petros and offer five Drats in payment?

How many minutes are there between 3.40pm and 4.25pm?

Digit card problem

Each of the digit cards has a home.
Can you find where it is?

□ 6 − 8 = □

$$\begin{array}{r} 4\square \\ -1\square \\ \hline 24 \end{array} \qquad \begin{array}{r} 6\square \\ -3\square \\ \hline \square 7 \end{array} \qquad \begin{array}{r} \square 4 \\ -\ \square \\ \hline \square 7 \end{array}$$

Abacus problems

Look at the numbers shown on these abacuses

Which pairs of abacuses have. . .

the largest difference an odd difference

a difference of 20 the smallest difference

start 70 68 ??

Missing numbers in subtractions

28

Some algebra work is concerned with ‘missing numbers’. Try and find the missing numbers in the following activities.

Simple equations

Each of the equations has a missing number.
There is only one possible answer for each equation.

$$16 - 9 = \square \qquad 24 - \square = 13 \qquad \square - 5 = 17$$

??

Open equations

Sometimes equations can have several possible answers.
There are two missing numbers in each of these equations.
Find different pairs of numbers for each equation to make it true.

$$\square - \triangle = 3 \qquad \square - 4 = \triangle \qquad 12 - \square = \triangle$$

??

Subtraction sequences

Find the missing numbers in these subtraction sequences.

34

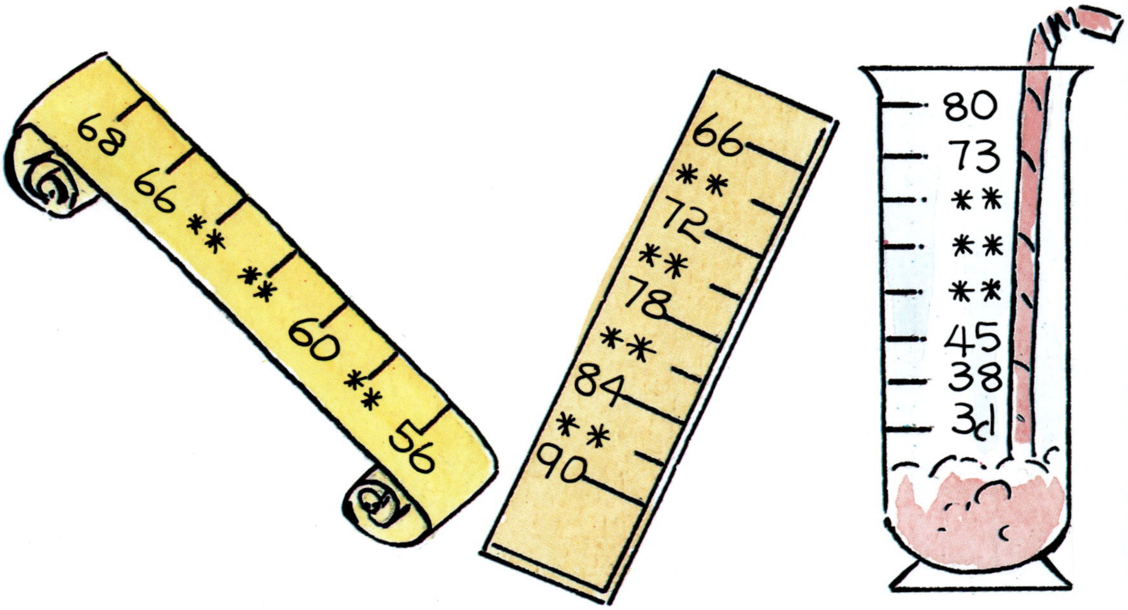

??

38 40

?? 60 ?? 56

Missing digits

Each star is a missing digit.
Use your subtraction skills to discover each missing digit.

8*	50	*2	7*	**
− 28	− **	− 56	− *3	− 25
*9	13	3*	60	56

54

Subtraction grids

Here are some subtraction grids with numbers missing.
Find the missing numbers.

??

−	8	9	10
12	4		
15			
20			

−		9	12
13	6		
			4
24			

−			
	9	8	7
	10	9	8
	11	10	9

Word problems

My difference is 5.
My total is 23.
Which two numbers am I?

Subtract 22 from me
then halve me.
The answer is 14.
What is my number?

??

Difference walls

Look carefully at this difference wall.
Can you see the number pattern?

48

Find missing numbers from these walls.

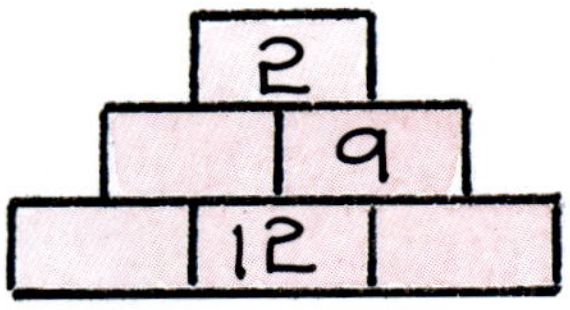

?? ?? ??

Number patterns and subtraction

Investigating and exploring number patterns is an important part of algebra.
Have fun exploring these number patterns.

Pyramid pattern

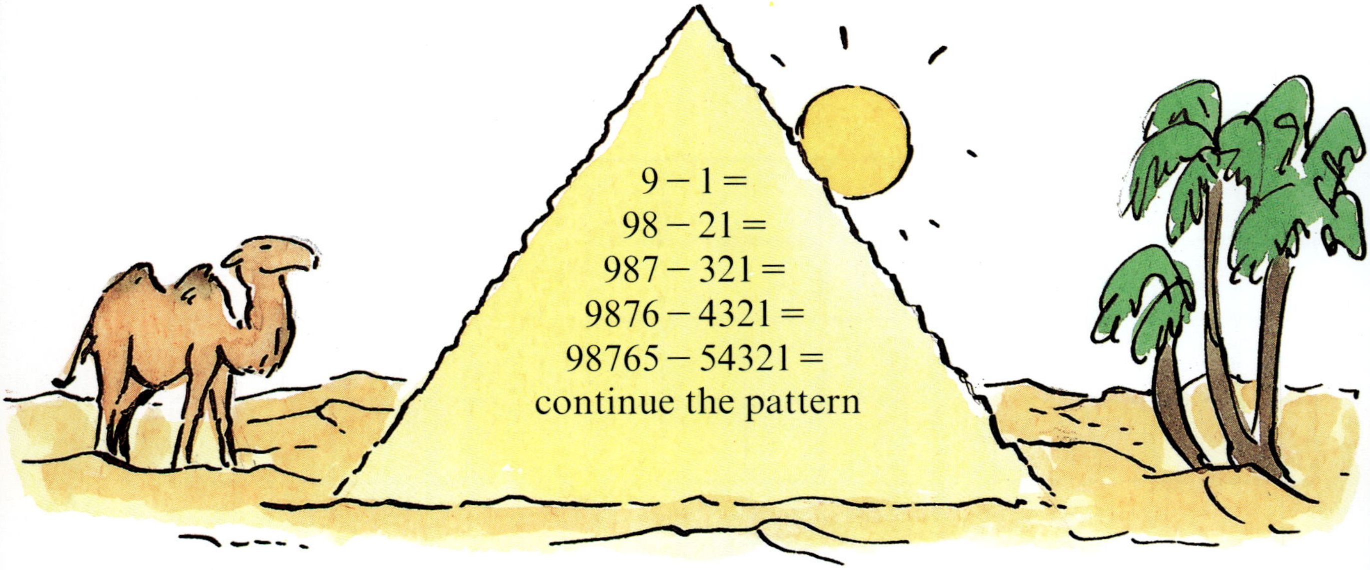

Square subtractions

To get a SQUARE NUMBER multiply two identical numbers together.
Here are some square numbers.

16 = (4 × 4)   36 = (6 × 6)   144 = (12 × 12)

Which two square numbers have been subtracted to give these answers?

11   40   45   96   48

Reversing digits

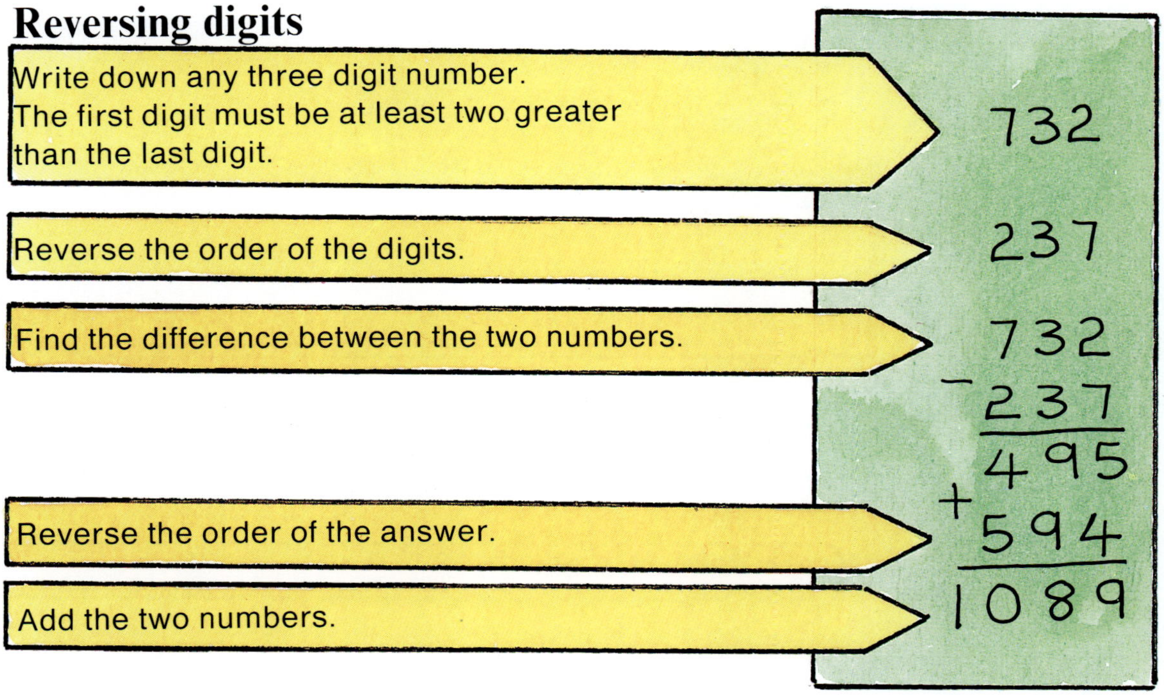

Repeat this with some three digit numbers of your own.
What do you notice?

Odds and evens

Which kind of number do you get if you subtract. . .

Subtracting with measures

Temperature drop

What is the temperature shown on these thermometers?

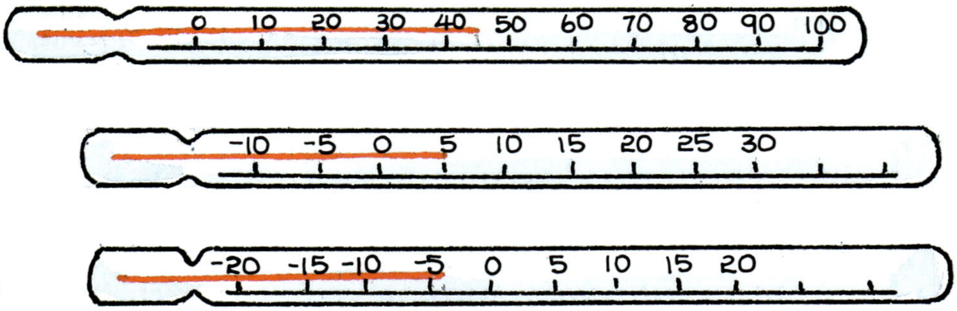

What will each reading be if the temperature drops by 9°C?

Height differences

How much taller is each of these animals than you?

Time check

Each of these clocks is 20 minutes fast.
What is the correct time for each clock?

Metric

Find the difference between these metric measures. . .

2.50m	3.25m
650g	175g
1.7 litres	2.5 litres
85cl	48cl
146cm	205cm
250ml	135ml

Imperial

In each of these pairs which is the smaller measure. . .

1 inch or 1 centimetre?

1 yard or 1 metre?

1 pound or 1 kilogram?

1 gallon or 1 litre?

Subtraction and data handling

Information is presented to us in all sorts of ways. . .

graphs tables lists charts Pictures maps diagrams

See if you can gather information from these sources.

	Apsley	Bingham	Coxhead	Dell	Exton
Bingham	4				
Coxhead	10	6			
Dell	15	11	5		
Exton	15	10	4	5	
Froggat	8	7	11	12	7

Distance in miles between villages

BUS TIMETABLE FROM APSLEY				
Apsley	0700	1115	1430	1800
Bingham	0720	1135	1450	1820
Coxhead	0750	1205	1520	1850
Dell	0810	1225	1540	1910
Exton	0830	1245	1600	1930
Froggat	0905	1320	1635	2005

- Which is the shortest route from Apsley to Dell?
- What is the difference in mileage going from Bingham to Exton by way of Froggat rather than through Coxhead?
- How many minutes does it take the bus to travel between Dell and Exton?
- How long does it take the bus to travel between Apsley and Coxhead?

MULTIPLYING

Multiplying

Multiplication squares

Learning your multiplication table facts can be hard work. Here are some activities which may help you understand and enjoy working with tables as well as remembering the answers.

Multiplication Square

X	2	3	4	5	6	7	8	9	10	11	12
2	4	6	8	10	12	14	16	18	20	22	24
3	6	9	12	15	18	21	24	27	30	33	36
4	8	12	16	20	24	28	32	36	40	44	48
5	10	15	20	25	30	35	40	45	50	55	60
6	12	18	24	30	36	42	48	54	60	66	72
7	14	21	28	35	42	49	56	63	70	77	84
8	16	24	32	40	48	56	64	72	80	88	96
9	18	27	36	45	54	63	72	81	90	99	108
10	20	30	40	50	60	70	80	90	100	110	120
11	22	33	44	55	66	77	88	99	110	121	132
12	24	36	48	60	72	84	96	108	120	132	144

Look at the answers in the green squares. What do you notice?

What do you notice about the answers in the orange, red and brown squares?

Looking for patterns helps you remember facts.

Table Parts

Here are some parts of the multiplication square. Can you work out the missing numbers?

	30
30	

	49	

9		

			64		

Mini Table Squares

Here are some mini table squares.
What are the missing numbers?

×	2	4	7
3			
5			
6			

×	3	7	
2			20
		42	
8			

×			
	25		
		36	
			49

Repeating Answers

Look at all the answers in the large multiplication square.
Some answers appear several times in the table.
Which answers only appear once in the table?

Multiplication build up

Repeated Addition

Multiplication is a quick way of working out repeated addition.

$5 + 5 + 5 + 5 + 5 + 5$ is the same as 5×6

$2 + 2 + 2$ is the same as 2×3

$13 + 13 + 13 + 13$ is the same as 13×4

Write a multiplication fact to show how many legs there are in each set of these strange insects .

Try to work out the answers to these problems without multiplying!

$9 \times 15 = 135$	$4 \times 23 = 92$	$3 \times 18 = 54$	$14 \times 20 = 280$
$9 \times 16 = ?$	$4 \times 25 = ?$	$3 \times 17 = ?$	$14 \times 21 = ?$

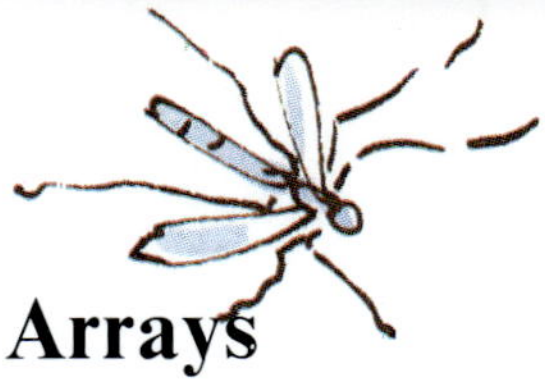

Arrays

Sometimes multiplication is shown in an **array** pattern.
Here are some array patterns.

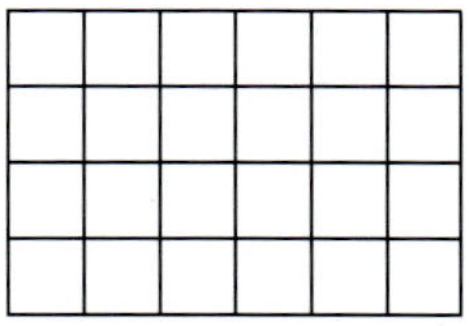

$4 \times 6 = 24$

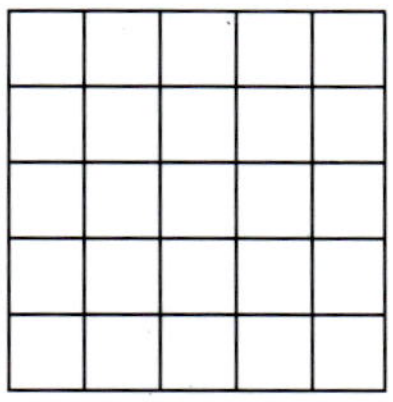

$5 \times 5 = 25$

$3 \times 11 = 33$

How many beetles in these arrays?

Here are some array patterns for 12.

Can you draw another?

Draw array patterns for...

Which numbers make square arrays?
Which numbers only make one array?

Multiplication puzzles

Use your multiplication skills to solve these puzzles.

Table Code

Find the answers to each table fact.
Use the code to change answers to letters.
Rearrange the letters to find the names of fruit.

CODE	A	C	E	G	H	L	M	N	O	P	R
	36	49	24	72	35	54	81	63	56	30	64

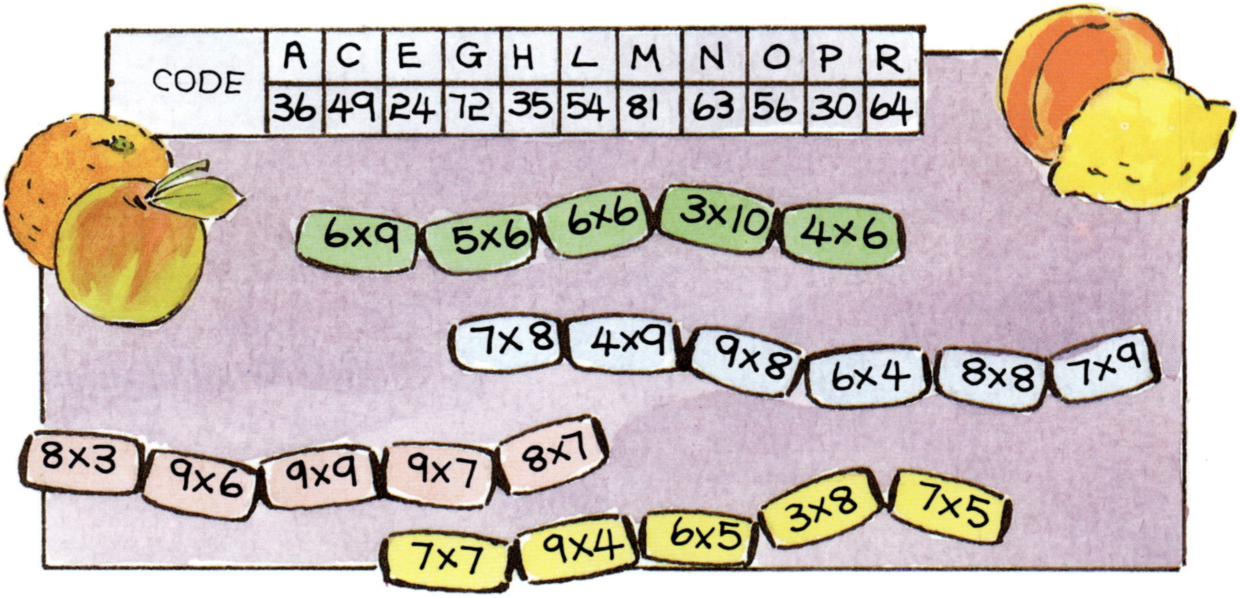

Multiple Honeycomb

The bee can only go into cells which are part of the 6 × table.
Trace routes through the honeycomb.
How many can you find?

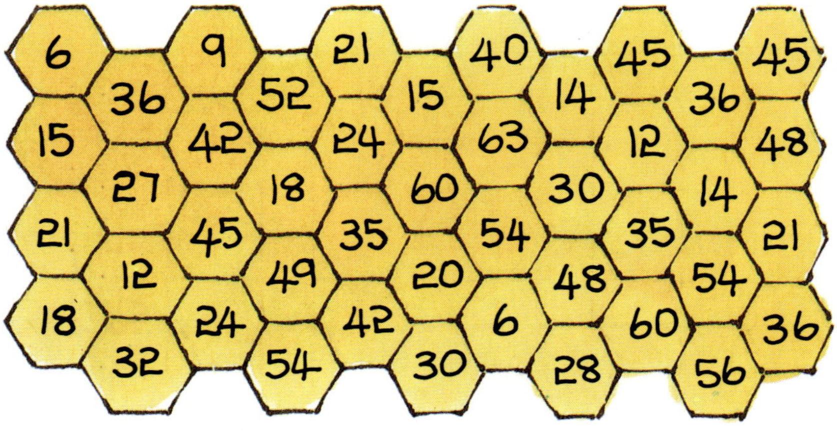

Calculator Puzzle

Calculate 35 × 23.
Turn the calculator upside down to see it cry!

Which multiplications will make these words when the calculator is turned upside down?

BIB BEE HOE SHE

Three Tables Puzzles

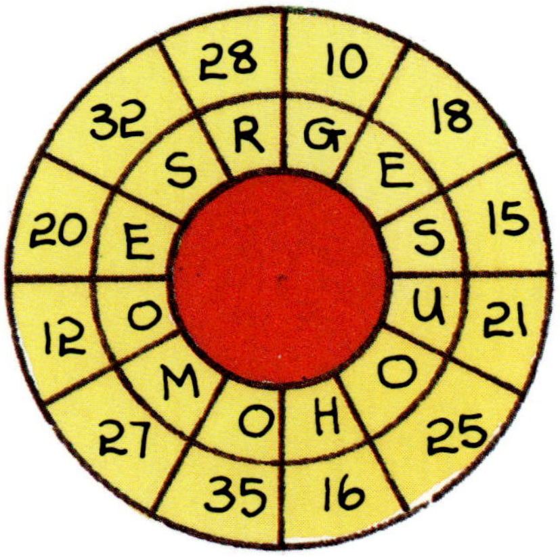

Find the numbers which are part of the 3 × table.
Write letters which go with the numbers.

Do the same for the 4 × and 5 × tables.

Rearrange the sets of letters to find the names of three farmyard animals.

Table Answer Puzzle

Think of the answers as words and spell them out using the blank spaces.
Find the two mystery animals.

2 × 4 = _ _ _ [_] _
4 × 3 = _ _ [_] _ _ _
5 × 4 = _ _ _ [_] _ _

4 × 8 = _ [_] _ _ _ _/_ _ _
6 × 7 = _ [_] _ _ _/_ _ _
7 × 9 = _ _ _ _ _/_ _ [_] _ _
6 × 6 = _ _ _ _ _ _/[_] _ _
9 × 9 = [_] _ _ _ _ _/_ _ _

Investigating with multiplication

A square number is a result of multiplying a number by itself.
Here are some square numbers.

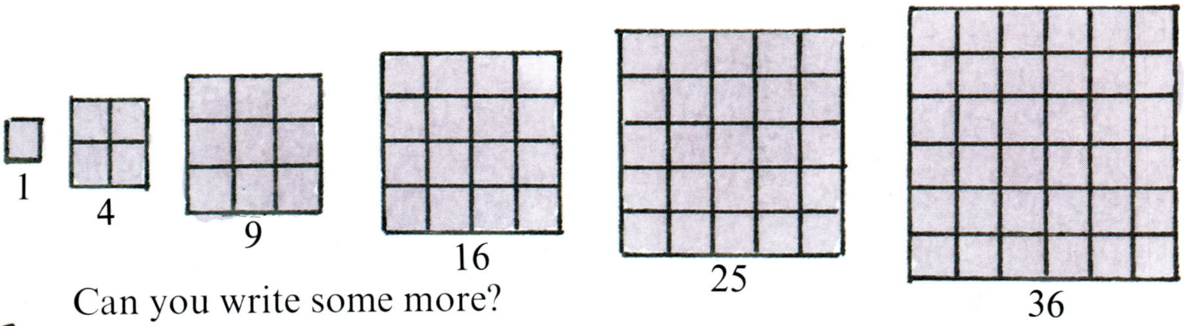

Can you write some more?

Square Numbers Investigation

Use a set of cards which have digits 0–9 written on them.

Arrange the cards to make one, two or three digit numbers.
Each number you make must be a square number.

Can you use up all the cards?
How many different square numbers can you make?

Calculator Investigation

Use the digit cards 0–9.
Choose any five cards and arrange them like this

Which arrangement will give the largest possible answer?

Try with different sets of numbers.

Digital roots are found by adding together the digits of a number.

36 → 3+6 → 9 42 → 4+2 → 6

Sometimes two additions are necessary to find the digital root.

76 → 7+6 → 13 → 1+3 → 4

Digital Root Investigation

Here are the digital roots of the × 3 and × 9 tables.

	digital root
$1 \times 3 = 3$	*3*
$2 \times 3 = 6$	*6*
$3 \times 3 = 9$	*9*
$4 \times 3 = 12$	*3*
$5 \times 3 = 15$	*6*
$6 \times 3 = 18$	*9*
$7 \times 3 = 21$	*3*
$8 \times 3 = 24$	*6*
$9 \times 3 = 27$	*9*
$10 \times 3 = 30$	*3*

	digital root
$1 \times 9 = 9$	*9*
$2 \times 9 = 18$	*9*
$3 \times 9 = 27$	*9*
$4 \times 9 = 36$	*9*
$5 \times 9 = 45$	*9*
$6 \times 9 = 54$	*9*
$7 \times 9 = 63$	*9*
$8 \times 9 = 72$	*9*
$9 \times 9 = 81$	*9*
$10 \times 9 = 90$	*9*

Investigate the digital roots of other tables.

Table Answers Investigation

The only answer in the nineties is 90.

90 — 9×10, 10×9

The only answer in the eighties is 80 and 81.

80 — 8×10, 10×8

Investigate this further.

81 — 9×9

Multiplication patterns

Multiplication can produce some interesting patterns for you to explore and investigate.

Calculator Pattern

- Start with 37037 on the display.
- Multiply it by 3 and find the digit which is repeated six times.

- Start with 37037 again.
- Multiply it by 5 and find which set of three digits is repeated.

Now you explore multiplying 37037 by any number between 3 and 27 to see what happens.

Table Patterns

Here is a hundred square.

1	2	3	4	5	6	7	8	9	10
11	12	13	14	15	16	17	18	19	20
21	22	23	24	25	26	27	28	29	30
31	32	33	34	35	36	37	38	39	40
41	42	43	44	45	46	47	48	49	50
51	52	53	54	55	56	57	58	59	60
61	62	63	64	65	66	67	68	69	70
71	72	73	74	75	76	77	78	79	80
81	82	83	84	85	86	87	88	89	90
91	92	93	94	95	96	97	98	99	100

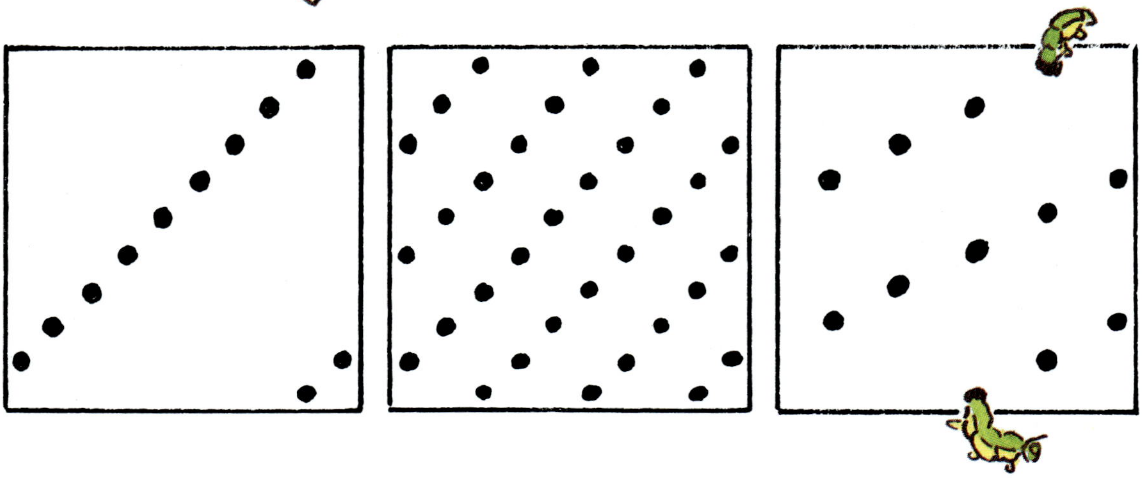

Multiples Patterns

The multiples of **9** are: 9 18 27 36 45 54 63...
Look at the last digit of each number.
What do you notice?

The multiples of **5** are: 5 10 15 20 25 30 35...
Look at the last digit of each number.
What do you notice?

> Investigate the pattern made by the last digit of other multiples.

Special Number Patterns

Multiply 37 by multiples of 3

37	37	37	37
× 3	× 6	× 9	× 12

Try other multiples of three.

Why is **37** special?

Is **74** special if you multiply it by multiples of 3?

Commutative Patterns

Multiplication is commutative because the order in which you multiply does not change the answer.

$3 \times 4 = 4 \times 3$

$6 \times 8 = 8 \times 6$

What are the missing numbers?

$5 \times 3 = ? \times 5$

$7 \times 2 = 2 \times ?$

$? \times 4 = ? \times 5$

$? \times 9 = 9 \times 4$

$6 \times ? = 8 \times 6$

Multiplication missing numbers

In these problems some numbers are missing.
Sometimes there will be only one answer.
Other times there will be several possible answers.
Can you discover which is which?

All Square

□ × □ ⇨ 8		□ × □ ⇨ 24		□ × □ ⇨ 30			
×	×	×	×	×	×		
□ × □ ⇨ 18		□ × □ ⇨ 56		□ × □ ⇨ 36			
⇩	⇩	⇩	⇩	⇩	⇩		
6	24	21	64	45	24		

Equations

4 × □ = 36 6 × 5 = □ □ × 7 = 49 9 × □ = 72 □ × 8 = 64

Open Equations

□ × △ = 24 □ × △ = 15 □ × △ = 81 □ × △ = 36

Sequences

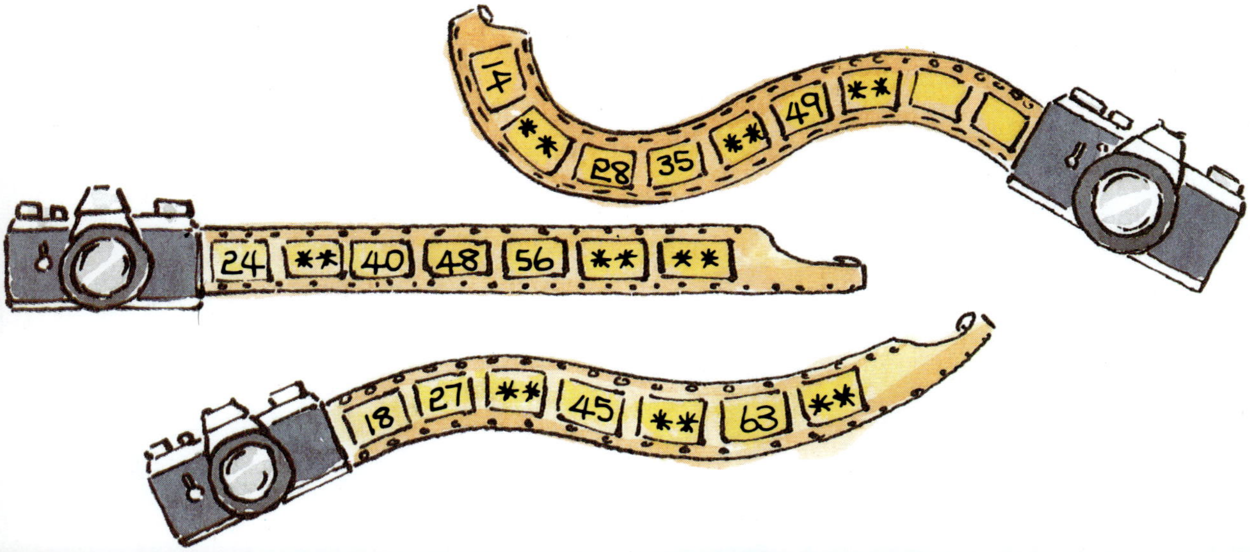

Missing Digits

Use digit cards 0–9.

Can you find a home for each of the digit cards?

□ × 8 = □□

□ × □ = □5

7 × 1□ = □0

□ × □ = 24

Functions Machines

Calculator multiplications

The calculator can be used to help you multiply large numbers. It can also be used to explore and experiment with multiplication of numbers.

Consecutive Numbers

Consecutive numbers are next-door neighbours like:

24 and 25 or 131 and 132

Find which consecutive numbers multiplied together make these answers.

☐ × ☐ = 210.　　☐ × ☐ = 342.

☐ × ☐ = 462.　　☐ × ☐ = 702.

Light Bars

5 × 8 = 40.

The answer of 40 is made from ten light bars.

4 × 8 = 32.

The answer of 32 is also made from ten light bars.

Can you find some multiplications which give answers made from ten light bars?

Forbidden Key

You are not allowed to touch the [2] key.
Here are two ways of finding the answer to 24×4 without touching the [2] key.

$8 \times 3 \times 4 =$
$6 \times 4 \times 4 =$

Can you find any more ways?

Try to find different ways of doing these multiplications without touching the [2] key.

Target

You can only use these number keys. [2] [3] [4] [6] [8]

You can only use these symbol keys. [×] [=]

Here is one way of hitting the target of 96.

[4] [×] [4] [×] [6] [=] 96.

How many different ways can you find of hitting the target of 96?

Multiplying and large numbers

Decade Times

Sometimes multiplying large numbers is just as easy as multiplying small numbers.

$4 \times 3 = 12$	$6 \times 8 = 48$
$40 \times 3 = 120$	$60 \times 8 = 480$

Try to answer these in your head.

Repeats

12345679 is a very large number!
It uses all the digits except 8.
It is also a special number.
Multiply it by 9, or a multiple of 9 and see why.

15873 is quite a large number.
Multiply it by 7, or a multiple of 7.
Why is it a special number?

15873×7 15873×14 15873×21

Russian Multiplication

Many years ago Russian peasants used a method of multiplying which only used adding, halving and doubling.
This is how it worked to answer 46 × 35.

STEP 1

Halve left column and ignore any remainder

Double right hand column

46	×	35
23		70
11		140
5		280
2		560
1		1120

Stop at 1

STEP 2

Cross out all Even numbers on left column and....

Cross out the corresponding number in the right column

STEP 3

Add the numbers remaining in the right column

~~46~~	~~×~~	~~35~~
23		70
11		140
5		280
~~2~~		~~560~~
1		1120
		1610

Try this method with some long multiplications of your own.

Quick Thinking

Which of these numbers can be changed to 5000 using multiplication only?

25 30 16 75 48

Multiplication methods

There are many different methods of working out the answers to multiplication problems.
Try some of these methods.

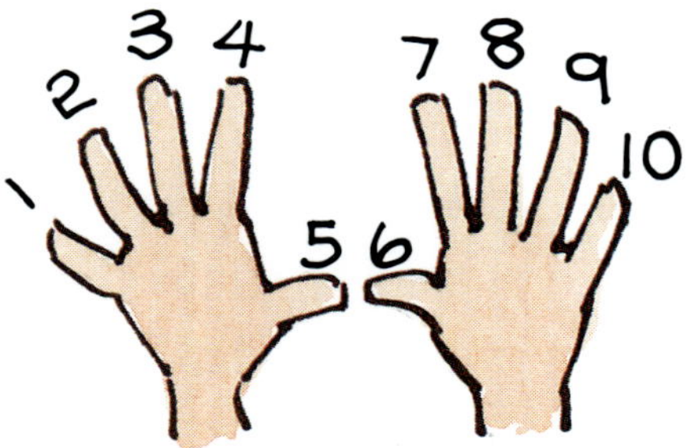

Finger Tables

Your fingers can be numbered like this. They can now be used to help you answer the 9 × table.

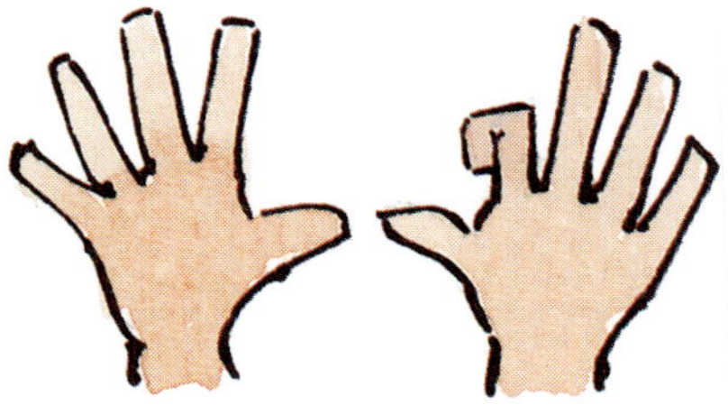

To find the answer to 9 × 7 hold down the seventh finger. There are six fingers to the left and three fingers to the right.

9 × 7 = 63

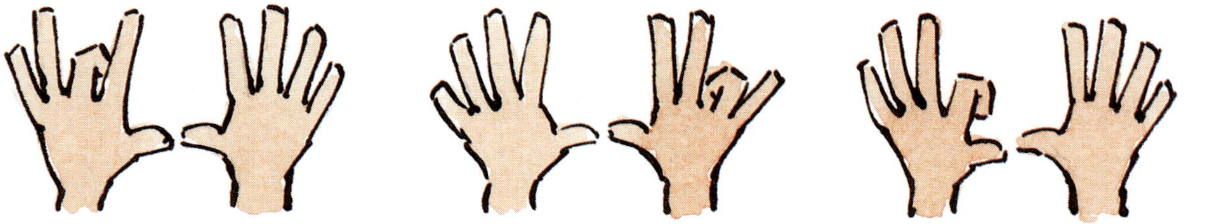

Which 9 × table fact do these hands show? Try some of your own.

Quick Elevens

Here is a quick way of multiplying by 11.

Try some 11× of your own

1 Write down the units digit.

2 Add each digit to its neighbour.

3 Write down the first digit.

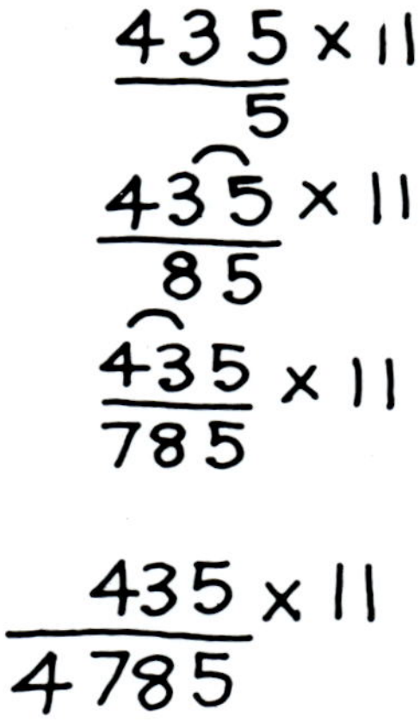

立 天 吉 心

Chinese Multiplication

Here is a way of doing long multiplication.

426 × 34

Draw a grid like this.

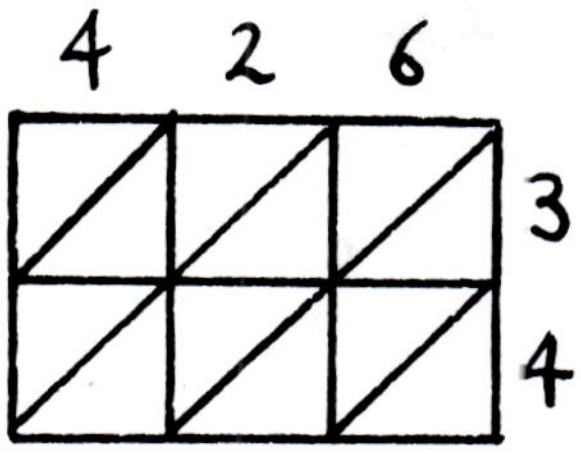

Fill in the grid by multiplying top and side numbers.

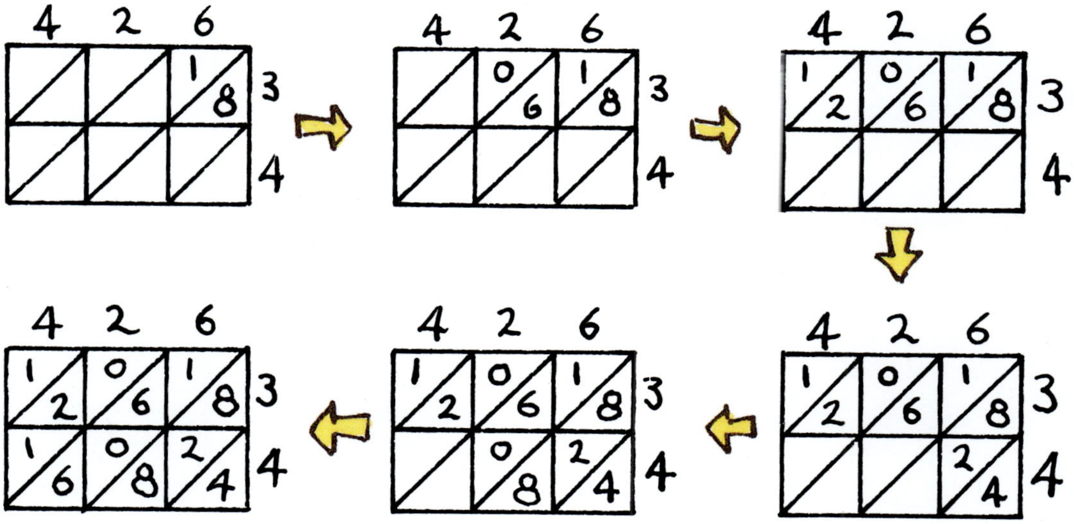

Now add the diagonal numbers, starting from the right.

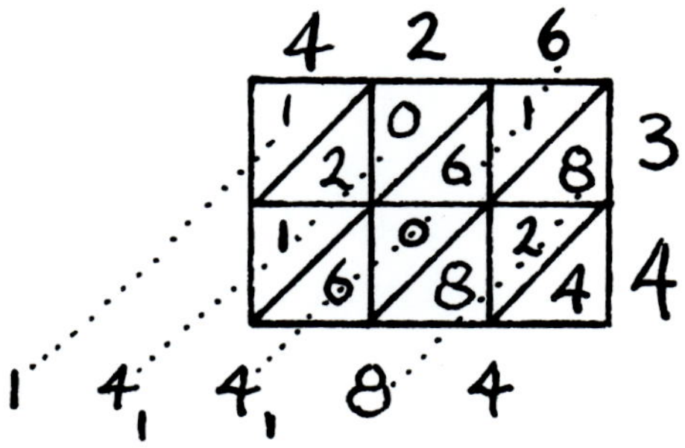

426 × 34 = 14484

Try these.

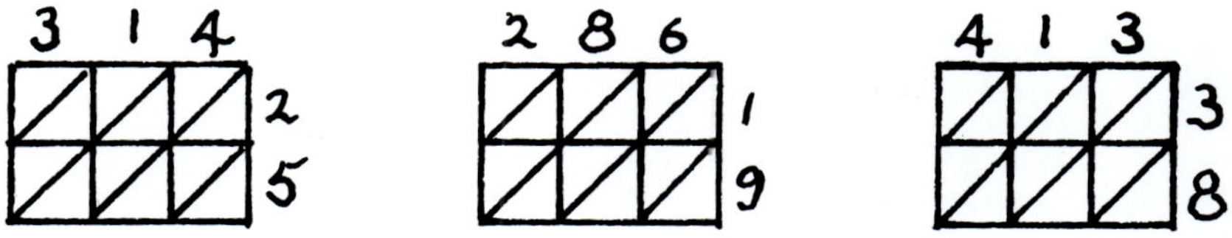

上 海 左

Multiplication games

Use your multiplication skills to play these games.

Patience

Use a deck of playing cards without the pictures.
Deal four cards face up and place the rest face down.
One card at a time is turned over from the deck.
This card can be placed below any of the face up cards.
When a column adds up to a multiple of 10 it can be removed.
Can you remove all the columns before the deck runs out?

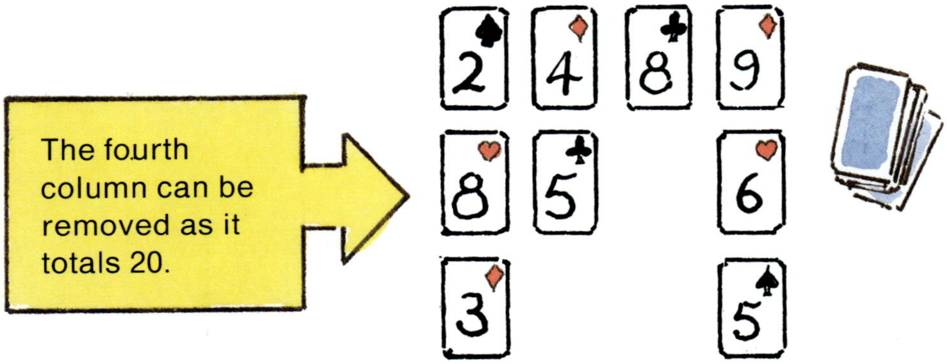

Multiples

3	6	12	6	24
18	12	9	12	18
6	12	6	36	12
30	18	6	12	18
18	6	18	15	6

A game for two players.
Take turns to roll a dice.
One player multiplies their score by 3 and covers the answer on the board.
The other player multiplies their score by 6 and covers the answer on the board.

Game 1: Try to cover the most numbers.
Game 2: Try to get three of your markers in a straight line.

Dominoes

Make a square with dominoes.
The total of each side
must be a multiple of 5.
How many squares can you make?
Can you make squares
whose side totals are multiples of 4?

Hex

Each player is trying to make a chain of their markers across the board.
The chain can be from side to side or top to bottom.
Multiply any two red digits together to cover an appropriate hexagon number.
A digit can be multiplied by itself.

Multiplication problems

Some multiplication problems are simple and straight forward, others are there to make you think.
See how you get on with these problems.

Operations Problem

You may use a calculator if you wish.

Digits which must be used. **2 4 6 8**

Operations which must be used × × +

4×2×8+6=70

Problems

What is the largest answer you can make?

What is the smallest answer you can make?

Can you make an answer of exactly 100?

Roman Problem

If you think multiplying is tricky think of the poor Roman children learning their tables!

Write out the 4× table as a Roman child would have done.

How would they have tackled this multiplication?

XIV
⊗ V

Numbers Problem

Use the digits **3 6 8** and the × sign.
Make all the different multiplication problems you can with them.
Here are two to start you off:

3 × 6 × 8 **68 × 3**

There are five more for you to find.
Which gives the biggest answer?

Abacus Problem

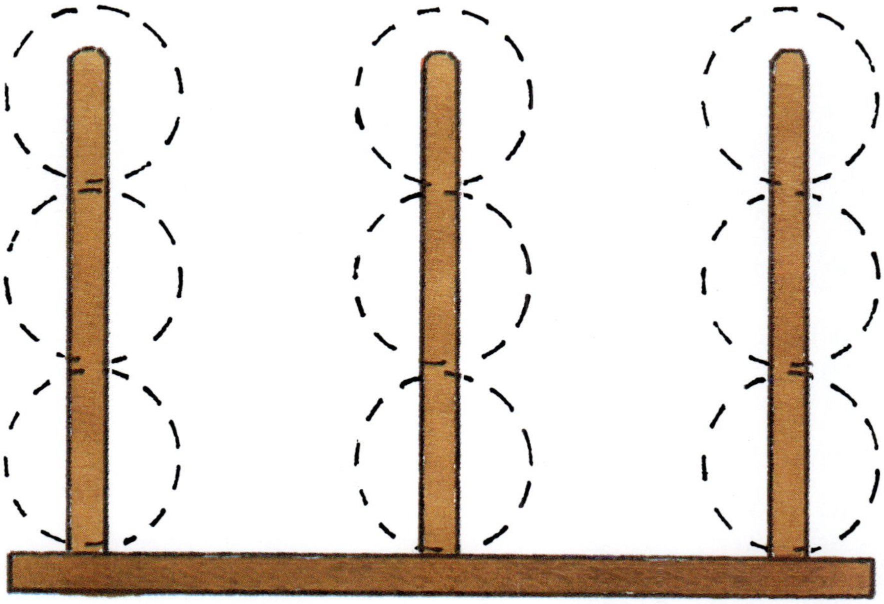

Place 3 small coins on the abacus to show numbers.

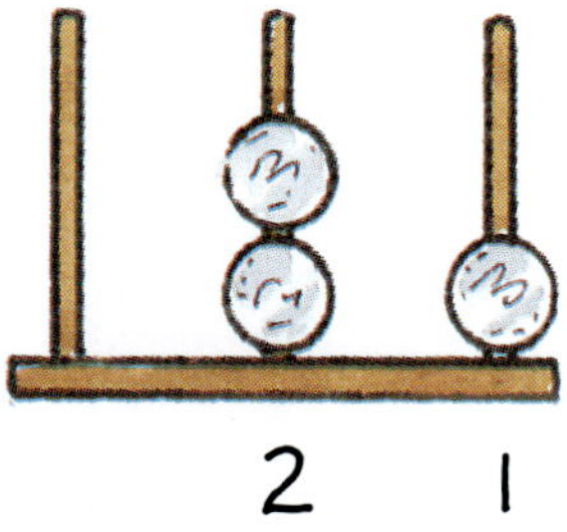

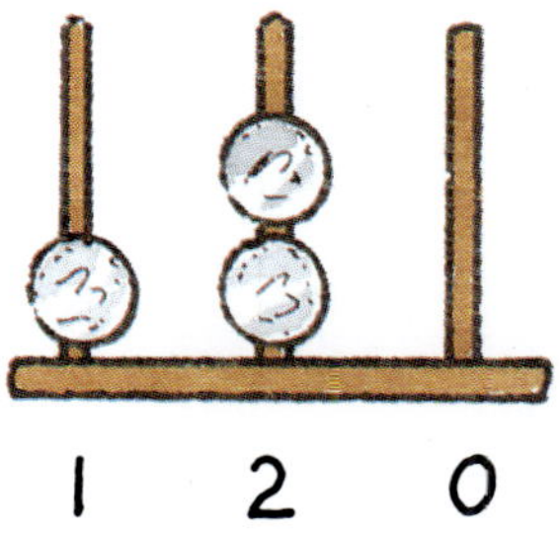

How many multiples of 3 can you show?
Can you make a number which is not a multiple of 3?

Measures and multiplying

Multiplication skills are needed when we use measures.

Finding Costs

What would a full box cost?

Calculating Time

How many minutes would each tape last?

Finding Areas

Measure and find the area of these shapes.

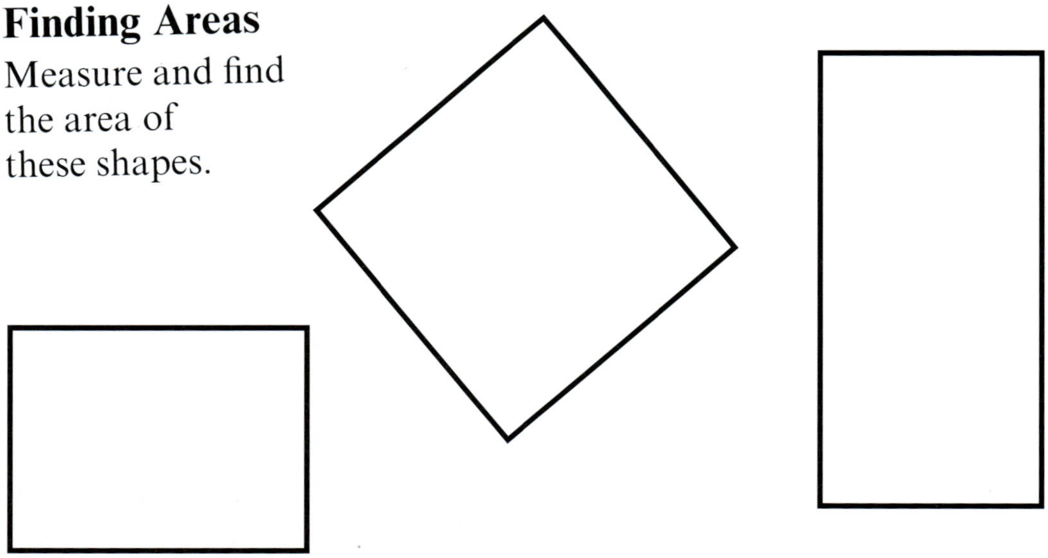

Calculating Weight

What will be the weight of 6 packets?

Calculating Quantity

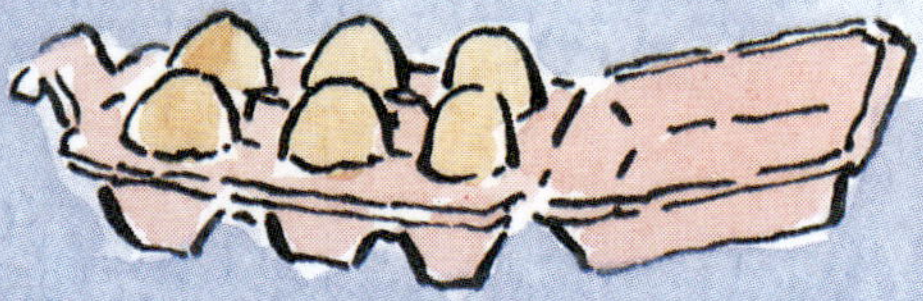

How many eggs in 8 boxes?

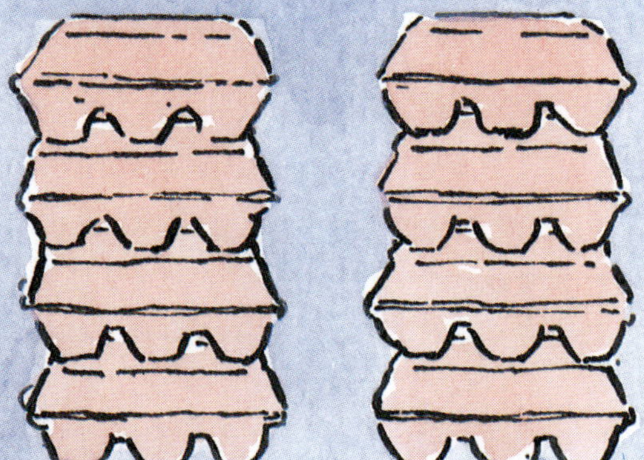

Calculating Capacity

What is the total capacity of 10 bottles?

Calculating Length

Each fence panel is 90cm long.
How far would seven such panels stretch?

Multiplication and data handling

Information is presented to us in all sorts of ways:

Use your skills to find the answers to these Venn diagram problems.

Use the numbers 1 to 30. Copy the Venn diagrams and write where you think the numbers 1–30 go on each one.

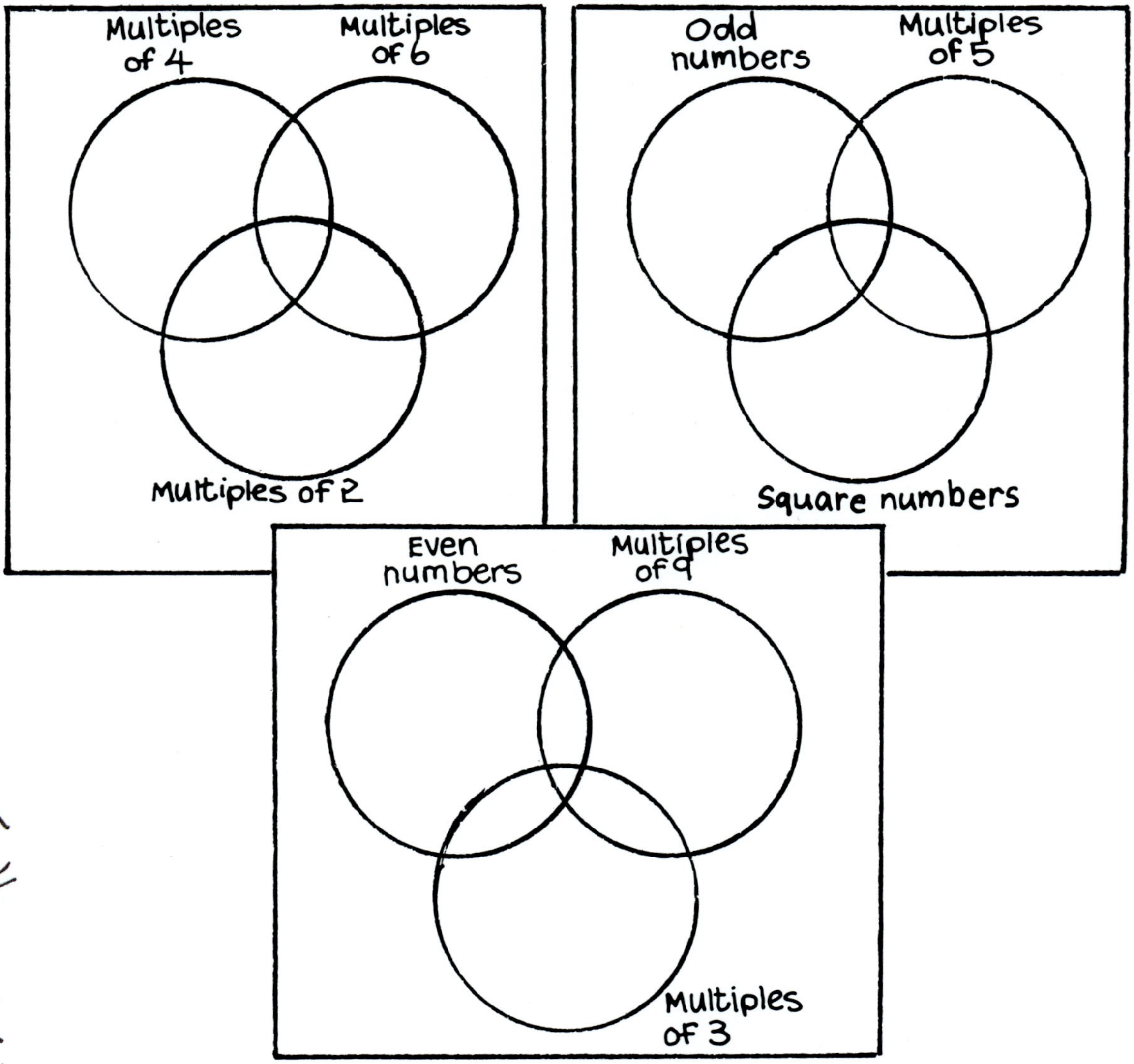

DIVIDING

Dividing

Division build up

There are lots of different ways of working out division problems. Here are two ways.

Repeated subtraction

Division is a quick way of working out the answers to repeated subtraction problems.

$12 \div 4$ *can mean:* how many times can you subtract four from 12?
or: how many fours in 12?

$12-4-4-4=0$ There are 3 fours in 12.

Try these repeated subtraction problems.

There are 27 segments in this giant orange.
How many children can have 3 segments each?

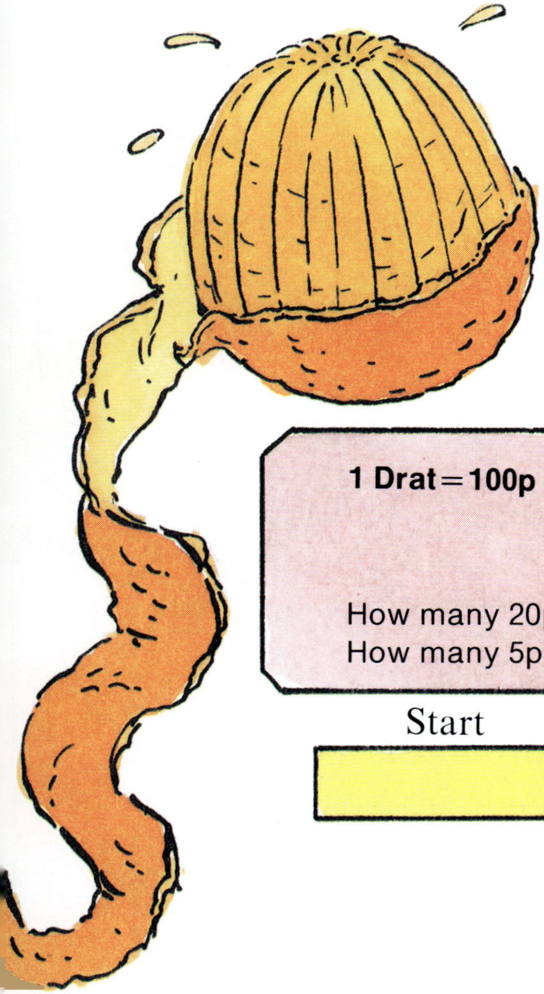

1 Drat = 100p

How many 20p coins are there in 1 Drat?
How many 5p coins are there in 1 Drat?

How many 14's have been taken away?

Equal sharing

Division is also a way of sharing equally.

12 ÷ 4 can mean share 12 equally into four sets.

12 = ③ ③ ③ ③

The four sets will each contain 3.

Try these equal sharing problems.

There are 32 strawberries in the box and there are four children. How many strawberries will each child have?

Forty-eight shared into 3 sets.
Draw what would be in each set.

Division ways

There are lots of different ways of working out the answers to division problems.

Sometimes one method makes the division easier, other times another is easier.

Here are some different methods for you to explore.

You can use a calculator to check the answers.

Halving

$6\overline{)516}$

Not very good at dividing by 6?
Try halving both numbers.

$3\overline{)258}$

Now work out the answer.

Try halving both numbers on these problems.
Does it change the final answer?

$6\overline{)312}$ $14\overline{)322}$ $18\overline{)828}$ $14\overline{)490}$

Can you see how to change some long divisions into short division?

Sometimes you can halve and halve again.

$12\overline{)672}$ → halve → $6\overline{)336}$ → halve → $3\overline{)168}$

Now try these.

$8\overline{)472}$ $12\overline{)768}$ $16\overline{)464}$ $16\overline{)512}$

Doubling

$5\overline{)235}$

Dividing by 5 is not too hard but dividing by 10 is easier!
Try doubling both numbers.

$10\overline{)470}$

Does the answer change?

Use the doubling method on these divisions.

$5\overline{)125}$ $5\overline{)315}$ $5\overline{)425}$ $5\overline{)435}$

Subtracting

$16\overline{)512}$

Let us find out how many sixteens there are in 512.
All you need to know is how to subtract and what 10 sixteens are.

$10 \times 16 = 160$

Subtract 10 sixteens	512 − 160 = 352	10
Subtract 10 sixteens	− 160 = 192	10
Subtract 10 sixteens	− 160 = 32	10
Subtract 1 sixteen	− 16 = 16	1
Subtract 1 sixteen	− 16 = 0	1
		32

number of sixteens in 512

Try these. $14\overline{)574}$ $16\overline{)656}$ $23\overline{)529}$ $26\overline{)806}$ $21\overline{)1092}$

Remainders

Sometimes when we divide there is a remainder left over. Explore these division activities and learn more about remainders.

Remainders galore

Divide 2519 by each number from 1 to 10 in turn. See what happens to the remainders.

$$1\overline{)2519}\quad \text{answer: } 2519\text{r}0 \qquad\qquad 2\overline{)2519}\quad \text{answer: } 1259\text{r}1$$

Now try 5039 divided by numbers from 1 to 10 in turn.

Zero remainders

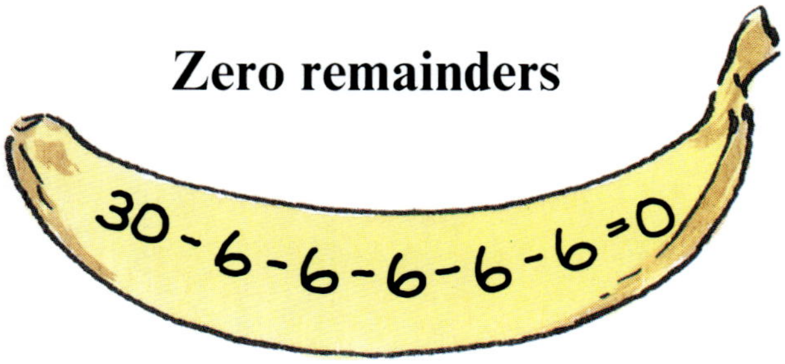

Start with 30.
Repeatedly subtract a single digit.
Which digits reach zero and have no remainder?

Try with different starting numbers.

Remainder puzzle

Divide each number by 8 to find the remainder.
Use the code to change remainders into letters.
Rearrange the letters to find the names of fruit.

CODE

Remainder	0	1	2	3	4	5	6	7
Letter	B	N	O	A	R	G	E	P

28	22	35	23

39	67	20	61	86

14	53	19	65	58	52

43	51	25	33	56	83

Remainder problem

Decimal remainders

Calculators do not give the remainders to division problems. They turn the remainder into a decimal.

$\begin{array}{r} 2r3 \\ 4\overline{)11} \end{array}$ $\qquad$ $\begin{array}{r} 2.75 \\ 4\overline{)11.00} \end{array}$

Try to change these remainders into decimals.

$\begin{array}{r} 11r1 \\ 2\overline{)23} \end{array}$ $\qquad$ $\begin{array}{r} \\ 2\overline{)23.0} \end{array}$

$\begin{array}{r} 6r2 \\ 5\overline{)32} \end{array}$ $\qquad$ $\begin{array}{r} 6. \\ 5\overline{)32.0} \end{array}$

$\begin{array}{r} 4r2 \\ 4\overline{)18} \end{array}$ $\qquad$ $\begin{array}{r} 4. \\ 4\overline{)18.0} \end{array}$

$\begin{array}{r} 8r1 \\ 4\overline{)33} \end{array}$ $\qquad$ $\begin{array}{r} 8. \\ 4\overline{)33.00} \end{array}$

Number search

I know a number which when divided by 2, 3, and 5 does not leave a remainder.
Can you discover my number?
Is there only one such number?

Division puzzles

Try these division puzzles. Knowing your tables will help!

Musical division

Work out each division fact. The answers tell you where to place the notes. Notes will be on a line or in a space.

The tune is a nursery rhyme. What do you think it could be?

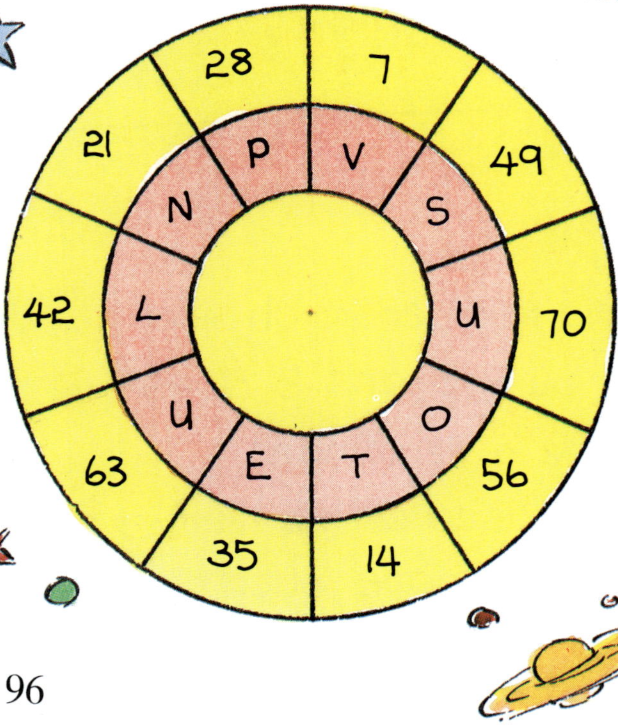

Division code

Divide each number in the outer circle by 7.
Write down the letters beside the EVEN answers.
Rearrange them to find a planet.

Write down the letters beside the ODD answers. Rearrange them to find another planet.

Division maze

Get the wasp to the apple.
It can only pass numbers which are divisible by **6** or **9**.
How many routes can you find?

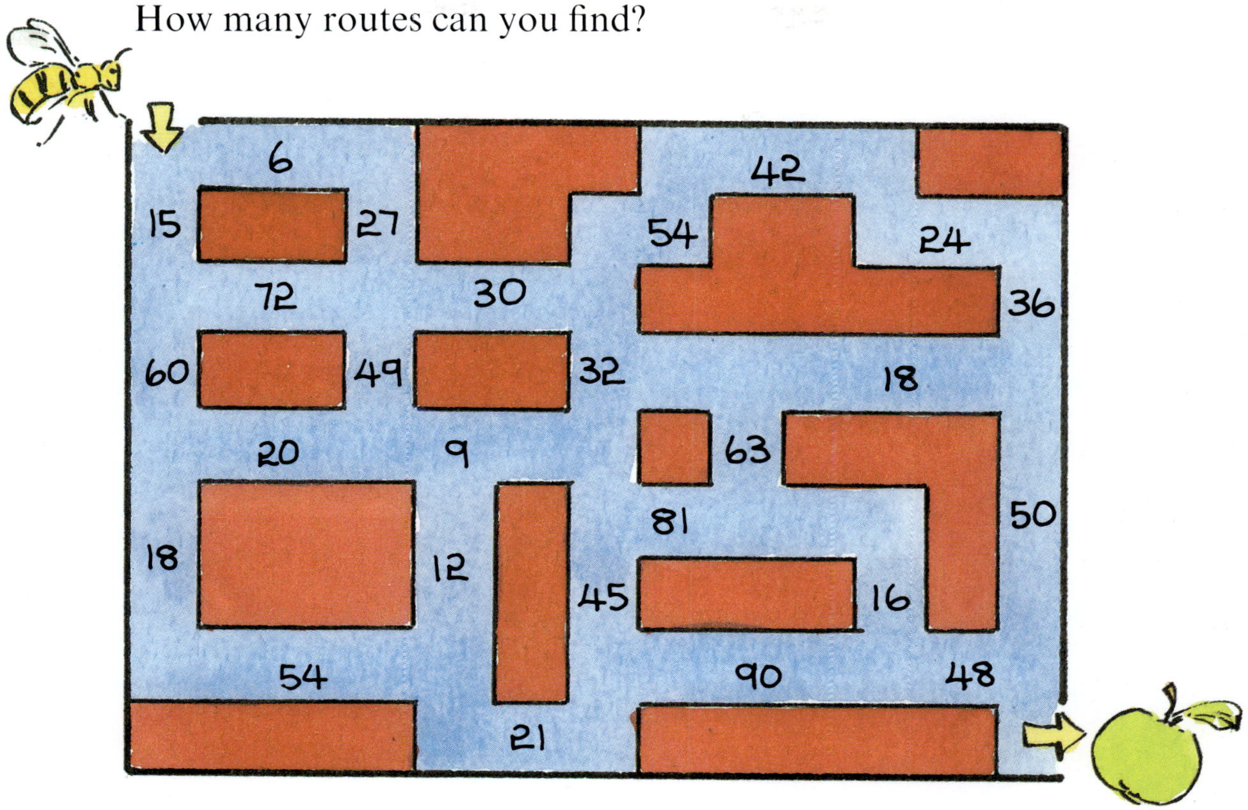

Division alphabet

Each letter of the alphabet has a number.

A	B	C	D	E	F	G	H	I	J	K	L	M	N	O	P	Q	R	S	T	U	V	W	X	Y	Z
1	2	3	4	5	6	7	8	9	10	11	12	13	14	15	16	17	18	19	20	21	22	23	24	25	26

Work out the answers to these division problems.
Find out the mystery words.

$\frac{36}{9}$ $\frac{90}{5}$ $\frac{126}{6}$ $\frac{52}{4}$

$\frac{72}{9}$ $\frac{135}{9}$ $\frac{54}{3}$ $\frac{98}{7}$

$\frac{60}{3}$ $\frac{72}{4}$ $\frac{36}{4}$ $\frac{10}{10}$ $\frac{112}{8}$ $\frac{49}{7}$ $\frac{84}{7}$ $\frac{25}{5}$

What have the words in common?

Missing numbers in division

In these problems some numbers are missing. Sometimes there will be only one answer. Othertimes there will be several possible answers. Can you discover which is which?

Equations

72 ÷ 9 = □ 64 ÷ □ = 8 □ ÷ 7 = 9
54 ÷ □ = 6 32 ÷ 8 = □ □ ÷ 6 = 7

Open equations

24 ÷ □ = △ □ ÷ △ = 3 36 ÷ □ = △
□ ÷ △ = 5 21 ÷ △ = □ □ ÷ △ = 8

Missing digits

Each star is a missing digit.
Can you work out what it should be?

31	41	*9	11*
*)124	8)3*8	5)4*5	*)702

Final digit

Each of these numbers is divisible by 9. The final digit of each number is missing. What should the digit be?

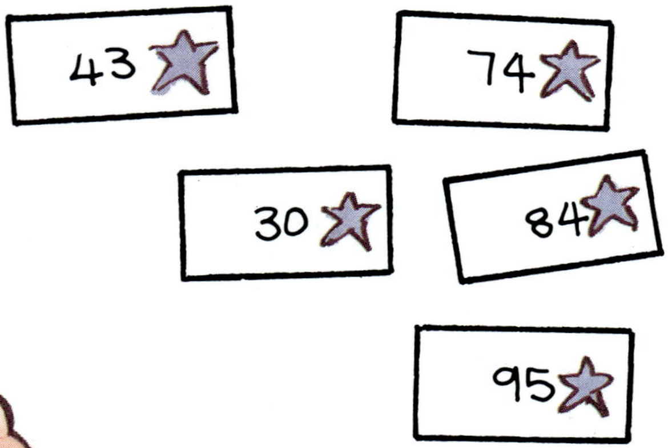

Function machines

Table grids

Copy and complete these table grids.

×		7
4	36	
		42

×		8
		72
10	50	

×		
	12	24
	28	56

Chains

What are the missing numbers in this "halving" chain?

What are the missing numbers in this "quartering" chain?

Division games

Use your division skills to play these games.

Fives and threes

Play dominoes in the usual way. As each domino is played total the numbers at each end. If the total is divisible by 5 or 3 the answer is the player's score. If the total is divisible by 5 AND 3 the score is the total of the two answers.

End total is 10
$10 \div 5 = 2$
Score is 2

End total is 15
$15 \div 5 = 3$ $15 \div 3 = 5$
Score is $3 + 5 = 8$

Division snap

Use a deck of cards without the pictures.
Divide the cards between the players.

The game is played like SNAP except you total the cards as they are played.

When the total is divisible by 2 you call SNAP and those cards are yours.

You can change the rules so that SNAP is called when the total is divisible by 3.

Division star

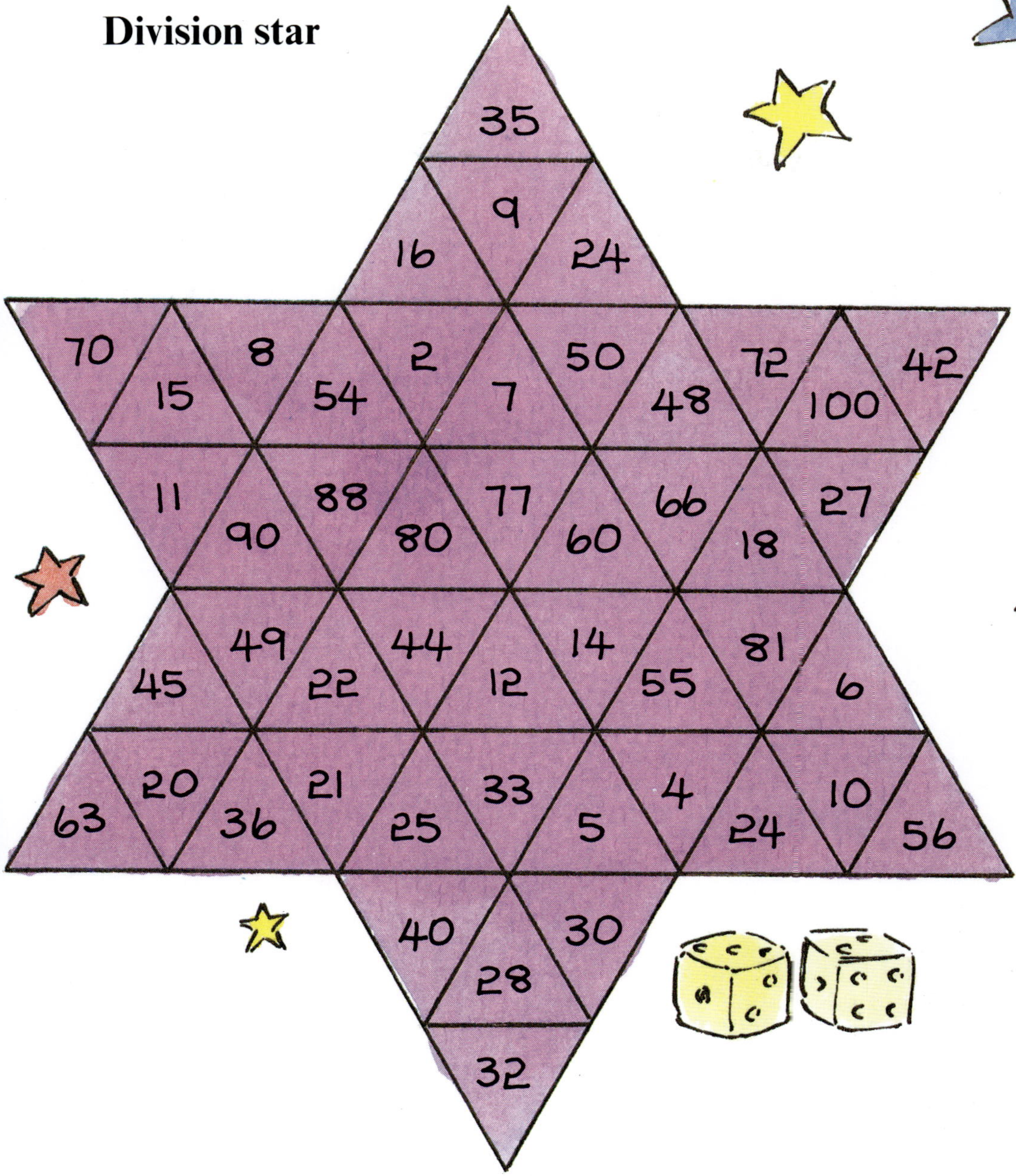

Roll 2 dice and total.
Cover any number on the star which this total will divide into exactly.

Game 1: First player to place 10 markers on the star wins.

Game 2: First player to have 4 touching markers wins.

Game 3: First player to complete a star point wins.

Game 4: The player who places most markers wins.

Investigating with division

Factor investigation

12 has six factors: : 1,2,3,4,6,12
17 has two factors : 1,17
9 has three factors : 1,3,9

Explore factors of numbers up to 50.

Which numbers have six factors?

Which numbers have only two factors?

Which numbers have an odd number of factors?

Digit card divisions

Use a set of digit cards numbered 0–9.

Use a division card. ÷

Make a set of four cards which give an exact answer.
No remainders allowed!
Here are two sets to start you off.

4 5 ÷ 9

3 4 ÷ 2

How many sets can you find?

Remainders

All these divisions give a remainder of 4.

9÷5 | 25÷7 | 28÷8 | 54÷10

Can you find other divisions which give a remainder of 4?
See if you can spot any patterns in your results.

Digit repeat

Here are the answers to some division problems.

6.6666666

2.2222222

11.111111

Which pairs of number could have been divided?
Find other parts of numbers which give answers of a repeating digit.

Forbidden key

You may use a calculator to answer these division problems . . . BUT . . . you must not touch the 5 key! Try to find different ways of doing each one.

5)90   9)135

5)165   4)252

Calculator divisions

A calculator can be used to help you divide large numbers. It can also be used to explore and experiment with division of numbers.

Double division

Enter a 3-digit number into your calculator.

536.

Make it a 6-digit number by repeating the same set of digits.

536536.

Divide by 7 . . .

Divide this answer by 11 . . .

Divide the answer by 13

What do you notice?

Try for other 6-digit numbers made in the same way.

Division estimator

An estimate is a "good guess" of the answer.
Choose the estimate you think is nearest the answer.

	ESTIMATES			ANSWERS
873÷9	0	50	100	
525÷5	100	150	200	
1757÷7	200	250	300	
1448÷4	300	350	400	
1488÷3	400	450	500	

Check your estimates by finding the answers.
Are you a good division estimator?

No remainders

You can only use these number keys.

[0] [2] [3] [5] [8]

You can only use these symbol keys.

[÷] [=]

Make up division problems which do not leave remainders, or give an answer with decimals.

[2] [0] [÷] [5] [=]   [5] [8] [÷] [2] [=]

How many can you find?

Calculator quickies

Here is the answer to a division problem.

8.

Can you make up ten division problems which give this answer?

Use your calculator to help you answer these problems.

437	365	156r*
6)****	**)4380	13)2037

Two-digit repeats

Start with 100 each time. Input one division operation. Try to make these answers.

9.0909090	1.5151515	1.8181818
3.0303030	4.5454545	1.0101010

Factors and divisibility

Factors are whole numbers which will exactly divide into other whole numbers.
The factors of **12** are **1, 2, 3, 4, 6, 12**.
The factors of **15** are **1, 3, 5, 15**.

Factor trees

Here are some factor trees

Can you see how each number is broken down into factors? If you multiply the numbers in each horizontal line you get the number you started with. Can you copy and complete these factor trees?

Factor query

Here are some factors of a number: 2 3 4
Which numbers, less than 50, could they be a factor of?

Divisible by 2

All even numbers are divisible by 2.
They have 2 as a factor.
Which of these numbers are divisible by 2?

46 93 124 205 438 522

Divisible by 5 and 10

Numbers which end in 0 are divisible by 10 and 5.
Numbers which end in 5 are divisible by 5.
Which of these numbers are divisible by 5?

95 120 254 506 435 600

Which of the above numbers are divisible by 10?

Divisible by 4

Which of these numbers are divisible by 4?

236 120 314 211 332 516

Look carefully at the last two digits of the numbers.
Can you find a rule for testing whether a number is divisible by 4?

Leap Year dates are divisible by 4. Which of these events happened on a leap year?

Division and prime numbers

Most whole numbers can be divided by another whole number without leaving a remainder.
A number which can only be divided by itself and one is called a PRIME NUMBER.
Explore these prime number activities.

Sieve of Eratosthenes

Eratosthenes was a Greek mathematician who lived 275–195 BC. He discovered the following method of finding prime numbers less than 100.
Copy this 100 grid.

1	2	3	4	5	6	7	8	9	10
11	12	13	14	15	16	17	18	19	20
21	22	23	24	25	26	27	28	29	30
31	32	33	34	35	36	37	38	39	40
41	42	43	44	45	46	47	48	49	50
51	52	53	54	55	56	57	58	59	60
61	62	63	64	65	66	67	68	69	70
71	72	73	74	75	76	77	78	79	80
81	82	83	84	85	86	87	88	89	90
91	92	93	94	95	96	97	98	99	100

Cross out 1.
Cross out all numbers divisible by 2, but not 2.
Cross out all numbers divisible by 3, but not 3.
Cross out all numbers divisible by 5, but not 5.
Cross out all numbers divisible by 7, but not 7.
Make a list of the numbers not crossed out.
These are prime numbers.

Domino primes

Here is a domino square.
The total of spots on each side is a prime number.
Can you make some prime number domino squares?

Card primes

Use a set of digit cards 0–9.

Arrange the cards to make 1-digit and 2-digit numbers.

Each number must be a prime number.
Can you use up all the cards?

Prime quiz

True or false?

Add one or subtract one from any prime number greater than 3.
Six will always divide into the answer.

In any consecutive twelve numbers under 100 there will always be at least two prime numbers.

There is always a prime number between consecutive square numbers up to 100.

Division, decimals and fractions

Decimals and fractions keep on popping up when you are exploring and investigating division.

Decimal divisions

A calculator turns a remainder into a decimal.

$25 \div 2 = 12.5$

Here are answers to five division problems. Find a pair of numbers which will divide to give each answer.

Which decimal remainders are possible if you divide by:

8   5   4   10

Repeating divisions

Some divisions do not work out exactly.
They go on, and on, and on . . .

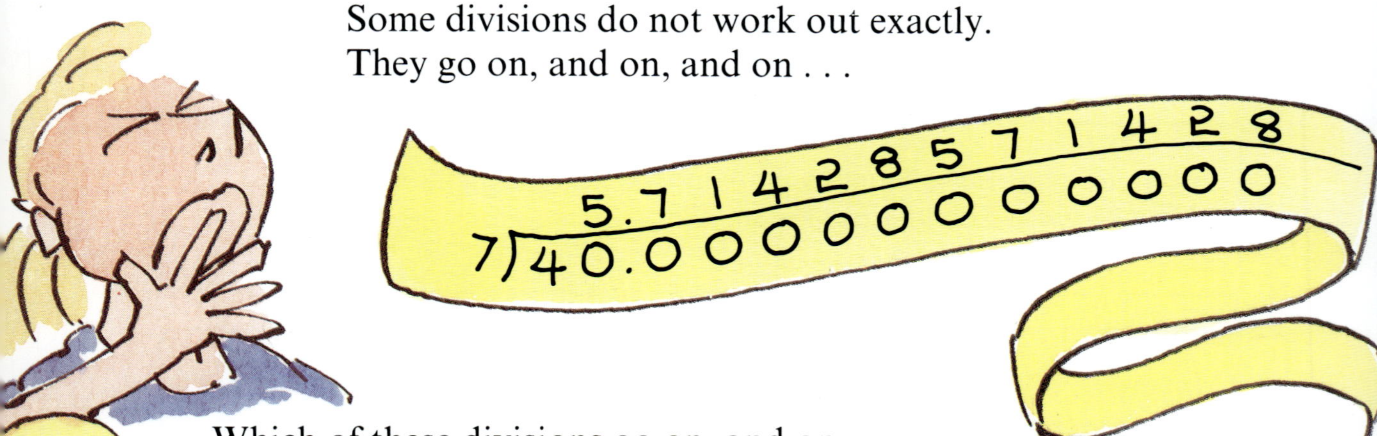

Which of these divisions go on, and on . . .

$7\overline{)36}$ $9\overline{)34}$ $5\overline{)28}$ $6\overline{)27}$ $3\overline{)41}$ $8\overline{)75}$

Can you discover which dividing numbers make divisions go on, and on . . .?

$\frac{22}{24}$ $\frac{100}{200}$

Dividing fractions

$\frac{1}{4}$

$\frac{3}{4}$ can mean $4\overline{)3.00}$ = 0.75 $\frac{2}{5}$ can mean $5\overline{)2.0}$ = 0.4

Can you change these fractions into decimals?

$\frac{5}{9}$

$\frac{3}{8}$ $\frac{4}{5}$ $\frac{1}{2}$ $\frac{7}{8}$ $\frac{3}{10}$ $\frac{1}{4}$

Nearly fractions

Some fractions do not change into decimals exactly.

$\frac{2}{3}$ ⇨ $3\overline{)2.000000}$ . . . . . . = 0.666666 . . . . . .

$\frac{2}{4}$

Which of these fractions do not change into decimals exactly?

$\frac{4}{7}$ $\frac{5}{6}$ $\frac{5}{8}$ $\frac{7}{9}$ $\frac{1}{3}$ $\frac{2}{5}$

Dice game

Each player needs a grid like this.

0									6

$\frac{6}{12}$

Players take turns to roll two dice to show two numbers. They decide in which order to divide the two numbers. The answer to the division is entered on the players' grid in one of the spaces. All the numbers on the grid must be in correct order. Numbers cannot be entered twice. Who can complete the grid first?

$3 \div 4 = 0.75$
or $4 \div 3 = 1.3333$

Measuring and division

Division skills can be used when we are measuring.

Length problem
Wall paper is usually sold in rolls which are 10 metres long.
Measure the height of your room.
How many lengths would you get from one roll?

Paper thickness
How thick do you think a sheet of paper is?
Measure a thick book.
Check the numbers of pages.
Can you calculate the thickness of one page?

Light weight
Can you find out how heavy one dried pea is?
Start by weighing lots of dried peas!
Try finding the weight of some other light objects.

Best buys

Which of these are the best buys?

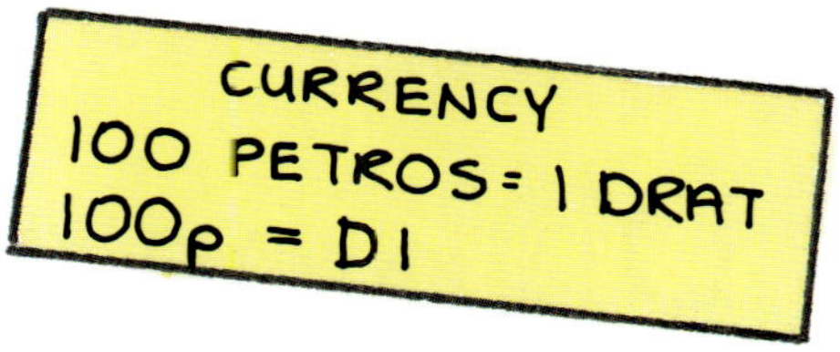

Part measures

Total each set. Find $\frac{1}{4}$ of each.

Liquid measures

A medicine spoon holds 5ml. How many doses will there be in each medicine bottle?

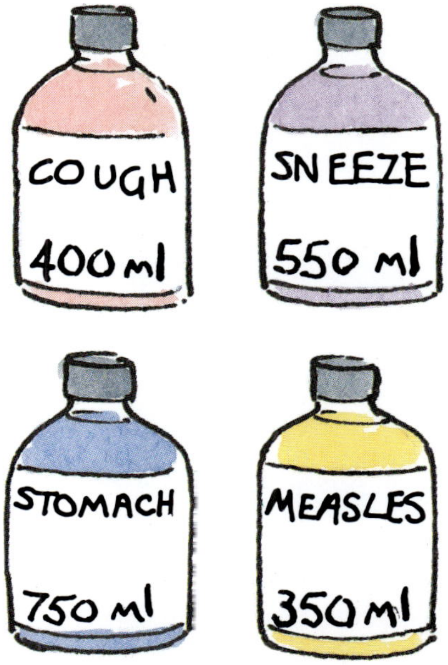

Division data handling

One way of showing information is with a pie chart.

This is a pie chart divided into the 24 hours of a day.

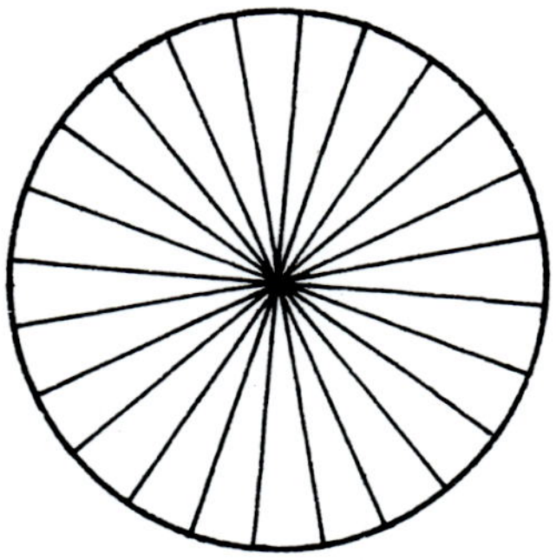

Here are pie charts showing how four children spent one Tuesday.

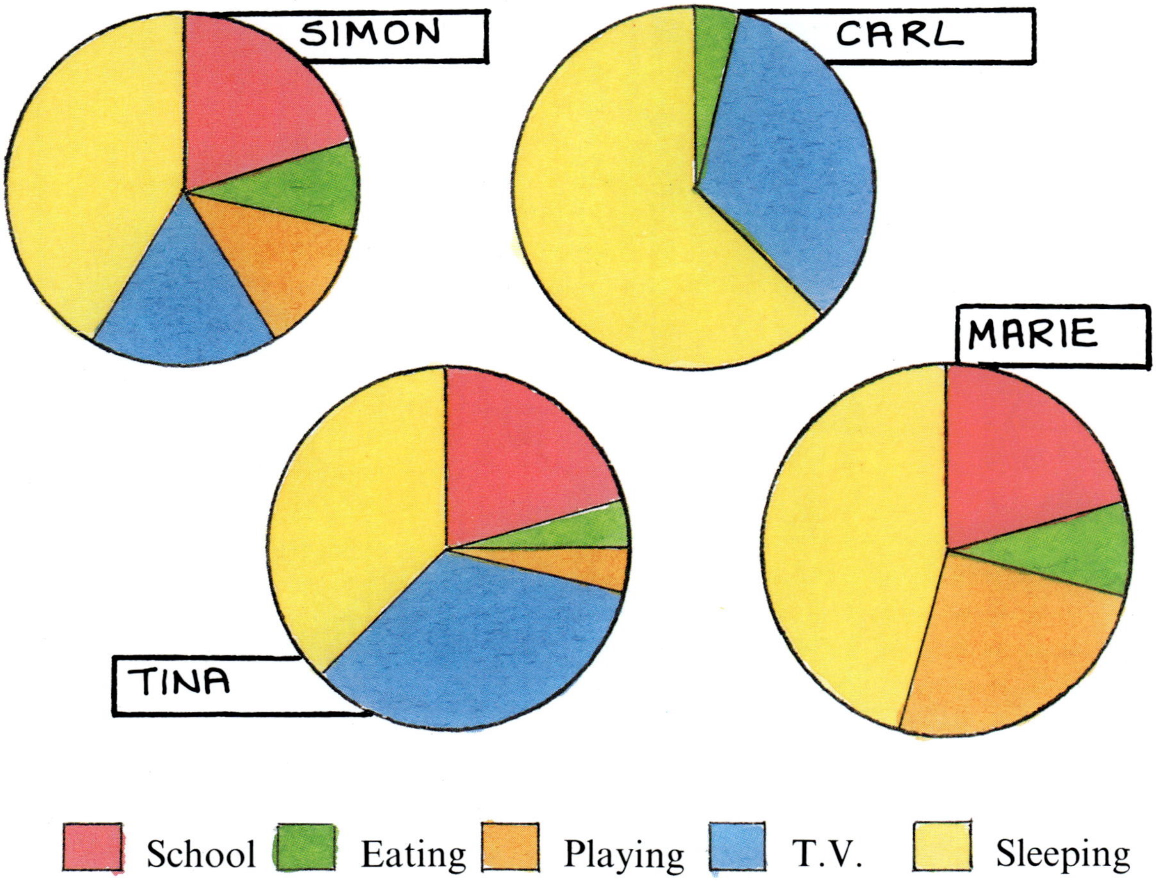

Who do you think was ill?

Who preferred to play than watch TV?

Who played and watched TV?

Make a pie chart to show how you spend a day

Glossary

Arrays

Patterns arranged in rows and columns.

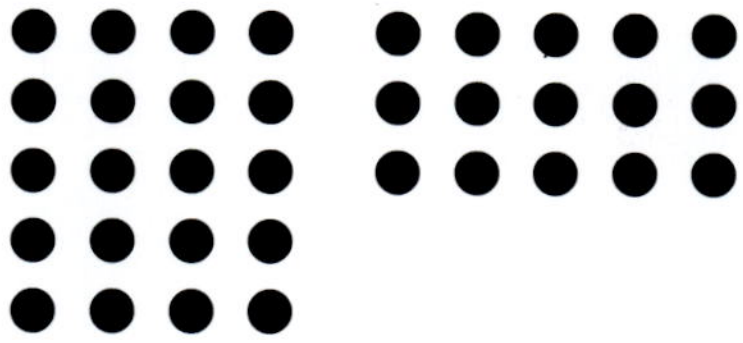

Commutative

Multiplication is commutative because the order in which you multiply it makes no difference to the answer.

$3 \times 6 = 6 \times 3$

Addition is also commutative.

$4 + 6 = 6 + 4$

Subtraction and division are not commutative.

Complementary addition

Calculating what must be added onto one number to make it the same as another number.

Consecutive numbers

These are numbers which follow on from each other e.g. 10, 11, 12 and 56, 57, 58, 59.

Difference

To find the difference between two numbers subtract them.
e.g. The difference between 20 and 26 is 6.

Digits

The digits are: 0, 1, 2, 3, 4, 5, 6, 7, 8, 9
Some numbers have two digits (35, 78, 95)
Some numbers have three digits (108, 567, 856).

Digital root

Digital root is the digit obtained by adding together the separate digits of a number.
24.. $(2+4=6)$.. digital root is 6.
67.. $(6+7=13)$.. $(1+3=4)$.. digital root is 4.

Estimation

An estimation is a sensible guess.

Even Numbers

Numbers which can be divided exactly by two.
Here are some even numbers: 2, 6, 18, 74, 90, 142.

Factors

Factors are whole numbers which will exactly divide into other whole numbers.
The factors of 8 are 1, 2, 4, 8
The factors of 20 are 1, 2, 4, 5, 10, 20

Imperial measures

Units used for measuring. The common ones still being used are:

Capacity – Pints (pts) Gallons (gals)
Length – Yards (yds) Feet (ft) Inches (ins)
Weight – Ounces (oz) Pounds (lb).

Magic squares

Numbers arranged in a square grid. Each row, column and diagonal total is the same.

16	3	2	13
5	10	11	8
9	6	7	12
4	15	14	1

Each row, column and diagonal totals 34.
It is a magic square.

Multiples

The multiples of 2 are 2 4 6 8 10 12...
The multiples of 3 are 3 6 9 12 15 18...
The multiples of 4 are 4 8 12 16 20 24...

Odd Numbers

Numbers which cannot be divided by two without leaving a remainder.
Here are some odd numbers: 3, 5, 27, 41, 89, 125.

Palindromes

Words and numbers which read the same when written forwards or backwards.
Here are some palindromes: MUM, DEED, 565, 2882

Perimeter

This is the distance measured all the way around a shape.

Prime Numbers

Prime numbers can only be divided by themselves and one. The prime numbers less than 100 are:
1 2 3 5 7 11 13 17 19 23 29 31 37 41 43 47 53 59 61 67 71 73 79 83 89 97

Square Numbers

A square number is obtained by multiplying a number by itself.
36 is a square number because $6 \times 6 = 36$
81 is a square number because $9 \times 9 = 81$.

Sum

To sum a set of numbers means to add them together. The sum of 12, 20 and 34 is 66.

Total

To total a set of numbers means to add them together. The total of 12, 20 and 34 is 66.

Answers

Page 6:

Magic squares

4	3	8
9	5	1
2	7	6

8	1	6
3	5	7
4	9	2

Totalling score
34

Page 7:

Addition walls

29
14 15
6 8 7
1 5 3 4

55
24 31
11 13 18
6 5 8 10
2 4 1 7 3

A possible answer

21
10 11
6 4 7
5 1 3 4

Page 10

Consecutive sums

You cannot make 2, 4, 8, 16, 32 by adding consecutive sums.
15 can be found in more than one way: $7+8$ and $1+2+3+4+5$.
Did you find any more?

Page 11

Make a table showing light bars needed to make each digit:

Digit	0	1	2	3	4	5	6	7	8	9
Light bars	6	2	5	5	4	5	6	4	7	6

Dice totals

With one dice each number should have the same chance of turning up. With two dice the total most likely to turn up is 7.

Page 12:

Alphabet code
Germany

Word sum puzzle

$$\begin{array}{r} 480 \\ +855 \\ \hline 1335 \end{array} \qquad \begin{array}{r} 8943 \\ +\ \ 53 \\ \hline 8996 \end{array}$$

Page 13:

Addition square puzzle

5	+	4	+	6	=	15
+		+		+		+
6	+	3	+	2	=	11
+		+		+		+
4	+	5	+	5	=	14
=		=		=		=
15	+	12	+	13	=	40

Word total puzzles
Bee is only worth 12.
Turkey is worth 100.
How did you do?

Page 15:

Consecutive numbers

$573=286+287$ $905=452+453$
$759=379+380$ $399=199+200$
$439=219+220$ $733=366+367$

All digit adds
There are many possible answers.
Here are 2 more:

$$\begin{array}{r} 243 \\ +675 \\ \hline 918 \end{array} \qquad \begin{array}{r} 782 \\ +154 \\ \hline 936 \end{array}$$

Page 16:

Addition grids

+	5	7	9	11
2	7	9	11	13
4	9	11	13	15
9	14	16	18	20
10	15	17	19	21
12	17	19	21	23

Problems grids
There are several answers. Here are two

6	1
2	3

5	2
3	2

Page 17:

Trangle grids
A possible answer

1
7 2 3
8 5 4
6

Page 18:

Dice problem
Opposite faces of a dice total 7.
16 spots; 23 spots; 23 spots.

Word problem
660 stamps

Mental problems
Half; three quarters; one whole

Page 19:

Digit problem
A possible answer

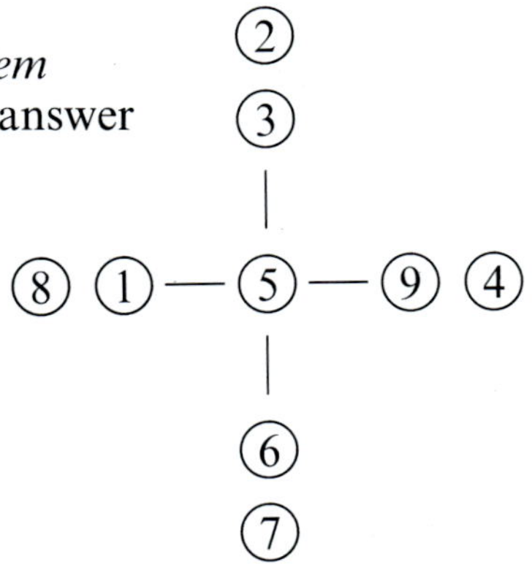

Page 22:

Simple equations
8 + 9 = 17   15 + 9 = 24   13 + 12 = 25

Open equations

0 + 10 = 10	0 + 3 = 3	7 + 0 + 13 + 20
1 + 9 = 10	1 + 3 = 4	7 + 1 + 12 = 20
2 + 8 = 10	2 + 3 = 5	7 + 2 + 11 = 20
3 + 7 = 10	3 + 3 = 6	7 + 3 + 10 = 20
⋮	⋮	⋮
10 + 0 = 10	Etc.	7 + 13 + 0 = 20

Additions sums

$$\begin{array}{r}43\\+\ 28\\\hline 71\end{array}\qquad\begin{array}{r}56\\+\ 98\\\hline 154\end{array}\qquad\begin{array}{r}38\\+\ 28\\\hline 66\end{array}$$

Page 23

Addition grids

+	4	5
3	7	8
6	10	11

Word problems
14 & 11
5

Addition sequence
45, 50, 55, 60, 65, 70, 75, 80
34, 38, 42, 46, 50, 54, 58, 62,
30, 37, 44, 51, 58, 65, 72, 79

Missing fractions
quarter
three-quarters
three-quarters

Page 24:

Triangle patterns
Each total is 1083676269

Palindromes
999
Most 3-digit numbers will result in palindromes when reversed and added. One exception is 196.

Page 25:

Odds and evens

0 + 0 = E	E + E = E	0 + E = 0
E + 0 = 0	E + E + E = E	0 + 0 + 0 = 0

Mental additons
79; 79; 90; 110; 138

Page 26:

Coin totals
All totals except 4p, 9p, 14p & 19p can be made.

Five coin totals
88p, 89p, 98p & 99p

Coin problem
The minimum is 3p, the maximum £1.50

Page 28:

How heavy?
1.800Kg (1800g); 650g; 280g

Finding perimeters
90cm; 140cm; 99cm; 108cm

Page 29:

Extra time
8·48; 12·10; 8·35; 4·55; 16·30

Metric addition
4.20m; 4.410Kg; 5.825 litre
12.4km; 152mm; 183ml

Imperial addition
26ins; 39oz; 18pts
34lb; 44ft; 44gal

Page 30:

Adding and data handling
28 children; 14 hours; 15 children (13 walk and 2 cycle) 17 children; 4 hours

Page 34:

Abacus numbers
022 030 018 009 088
Halving
123 107 71 69

Page 35:

Smaller measures
1.55m (155cm) 500ml 375g D3.55

Page 36:

Pairs
15p 90p 30p 95p

Page 37:

Digit cards
7 1 5
8 3 9
4 6 2

Difference challenge
There are more than 20 different answers.

Page 38:

All change
(10p, 50p, D1)
(2p, 2p, 20p, 50p, D1, D1, D1)
(1p, 2p, 5p, 20p, 20p)
(2p, 2p, 5p, D1, D1)
2p, 20p, D1, D1, D1)

Page 39:

Equations
7+8=15 3+11=14 5+13=18
26+6=32 14+23=37 28+22=50

Time check
30min 110min 65min 80min

Round up
13 68 14 46 62

Page 42:

Digit switch
The answer will always be the same.

Page 43:

Darts score
Claire: 250 Alan: 263

Consecutives
157 − 28 − 29 = 100
213 − 56 − 57 = 100
171 − 35 − 36 = 100
187 − 43 − 44 = 100
239 − 69 − 70 = 100

Page 44:
Word teaser
There are several possible answers e.g.

$$\begin{array}{r} 268 \\ -\ 94 \\ \hline 174 \end{array} \qquad \begin{array}{r} 9456 \\ -\ 72 \\ \hline 9384 \end{array}$$

Code wheel
Perth, Hague, Paris

Page 45:
Bubble puzzle
10 − 3 = 7   14 − 5 = 9
17 − 9 = 8   12 − 8 = 4   15 − 10 = 5

Mystery names
Toni   Steven

Page 50:
Bracket problem
(18 − 3) − 2   15 − (4 − 2)   24 − (3 + 8)
9 + (11 − 7)   18 − (8 − 5) − 2
16 − 2 − (9 − 8)

Mental problems
three quarters,   one quarter,   half

Word problems
155petros   45 minutes

Page 51:
Digit card problem
16 − 8 = 8

$$\begin{array}{r} 40 \\ -\ 16 \\ \hline 24 \end{array} \qquad \begin{array}{r} 69 \\ -\ 32 \\ \hline 37 \end{array} \quad \begin{array}{r} 54 \\ -\ 7 \\ \hline 47 \end{array}$$

Abaci problem
A&D, D and any other one, B&F, C&D

Page 52:
Simple equations
16 − 9 = 7   24 − 11 = 13   22 − 5 = 17

Open equations

3, 0	4, 0	0, 12
4, 1	5, 1	1, 11
5, 2	6, 2	2, 10
6, 3 etc	7, 3 etc	3, 9
		etc to
		12, 0

Subtraction sequences
68 66 64 62 60 58 56
90 87 84 81 78 75 72 69
80 73 66 59 52 45 38 31

Page 53:
Missing digits

$$\begin{array}{r} 87 \\ -\ 28 \\ \hline 59 \end{array} \quad \begin{array}{r} 50 \\ -\ 37 \\ \hline 13 \end{array} \quad \begin{array}{r} 92 \\ -\ 56 \\ \hline 36 \end{array} \quad \begin{array}{r} 73 \\ -\ 13 \\ \hline 60 \end{array} \quad \begin{array}{r} 81 \\ -\ 25 \\ \hline 56 \end{array}$$

Subtraction grids

−	8	9	10
12	4	3	2
15	7	6	5
20	12	11	10

−	7	9	12
13	6	4	1
16	9	7	4
24	17	15	12

−	5	6	7
14	9	8	7
15	10	9	8
16	11	10	9

Word problems
14, 9   50

Difference walls

4
6 10
8 2 12

2
7 9
5 12 3

1
2 1
2 4 3
6 8 12 15

Page 54:
Square subtractions
36–25   49–9   81–36   100–4   64–16

Page 55:
Reversing digits
The answer will always be 1089

Odds and evens
even, even, odd

Page 56:
45°C   5°C   −4°C
36°C   −4°C   −13°C

Page 57:
Time check
3.45   9.25   17.55   1.05   9.15   2.30

Metric
0.75m (75cm)   475g
0.8 litres   37cl   59cm   115ml

Imperial
cm   yard   pound   litre

Page 58:
Shortest route: Apsley, Bingham, Coxhead, Dell (15 miles); 4 miles; 20min; 50min

Page 63:
Table parts

25	30
30	36

42	49	56
48	56	64

9		
	16	
		25

40	48	56	64	72	80

Mini table squares

×	2	4	7
3	6	12	21
5	10	20	35
6	12	24	42

×	3	7	10
2	6	14	20
6	18	42	60
8	24	56	80

×	5	6	7
5	25	30	35
6	30	36	42
7	35	42	49

Repeating answers
4 9 25 49 64 81 100 121 144

Page 64:
Repeated addition
$8 \times 9 = 72$ $6 \times 7 = 42$ $4 \times 6 = 24$

$9 \times 15 = 135$ $4 \times 23 = 92$
$9 \times 16 = 144$ $4 \times 25 = 100$

$3 \times 18 = 54$ $14 \times 20 = 280$
$3 \times 17 = 51$ $14 \times 21 = 294$

Page 65:
Arrays
54   36   108
16 and 25 make square arrays.
25 only makes one array.

Page 66:
Table code
APPLE   ORANGE

LEMON (or MELON) PEACH

Multiple honeycomb
There are four routes

Page 67:
Three tables puzzles
MOUSE HORSE GOOSE

Table answer puzzle
HEN HORSE

Page 70:
Table patterns
9 × 3 × 8 ×

Commutative pattern
5 × 3 = 3 × 5 7 × 2 = 2 × 7

5 × 4 = 4 × 5 4 × 9 = 9 × 4

6 × 8 = 8 × 6

Page 72:
All square

2	4
3	6

3	8
7	8

5	6
9	4

Equations
4 × 9 = 36 6 × 5 = 30 7 × 7 = 49

9 × 8 = 72 8 × 8 = 64

Open equations
There are several answers to each equation.
24 = (1 × 24) (2 × 12) (3 × 8) (4 × 6) (6 × 4) (8 × 3) (12 × 2) (24 × 1)

15 = (1 × 15) (3 × 5) (5 × 3) (15 × 1)

81 = (1 × 81) (3 × 27) (9 × 9) (27 × 3) (81 × 1)

36 = (1 × 36) (2 × 18) (3 × 12) (4 × 9) (6 × 6) (9 × 4) (12 × 3) (18 × 2) (36 × 1)

Sequences
14 21 28 35 42 49 56 63 70
24 32 40 48 56 64 72
18 27 36 45 54 63 72

Page 73:
Missing digits
2 × 8 = 16 9 × 5 = 45

7 × 10 = 70 8 × 3 = 24

Function machine
out: 21 28 42 49 56 63
in: 4 5 6 8 9 10

Page 74:
Consecutive numbers
14 × 15 = 210 18 × 19 = 342

21 × 22 = 462 26 × 27 = 702

Light bars
There are many answers
(e.g. 8 × 8 7 × 10)

Page 77:
Quick thinking
25

Page 78:
Finger tables
3 × 9 9 × 9 4 × 9

Page 82:
Operations problem
Largest: 2 + 4 × 8 × 6 = 288

Smallest: 6 × 4 × 2 + 8 = 56

100 = 8 × 6 × 2 + 4

Page 83:
Numbers problem
3 × 6 × 8 36 × 8 63 × 8 86 × 3

68 × 3 83 × 6 38 × 6

The biggest answer is 63 × 8 = 504

Abacus problem
3 12 21 30 102 111 120 201 210 300
All the numbers will be multiples of 3.

Page 84:
Finding costs
288p (2.88 Drats)

Calculating time
120min 180min 240min

Finding areas
12cm^2 16cm^2 18cm^2

Page 85:
Calculating weight
2520g (2.520kg)

Calculating quantity
48

Calculating capacity
2500ml (2.5l)

Calculating length
630cm (6.3m)

Page 86:
Venn diagrams

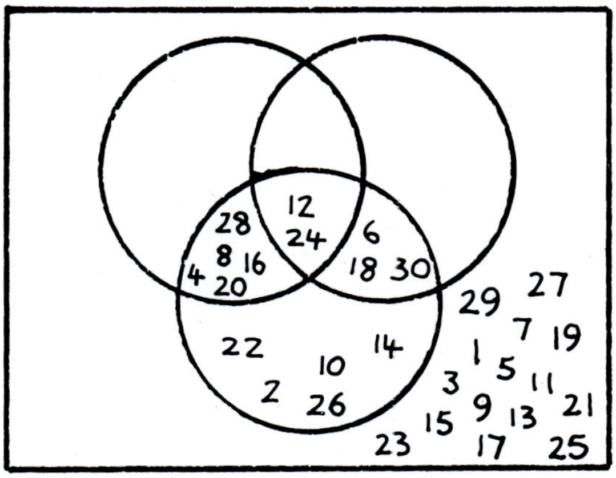

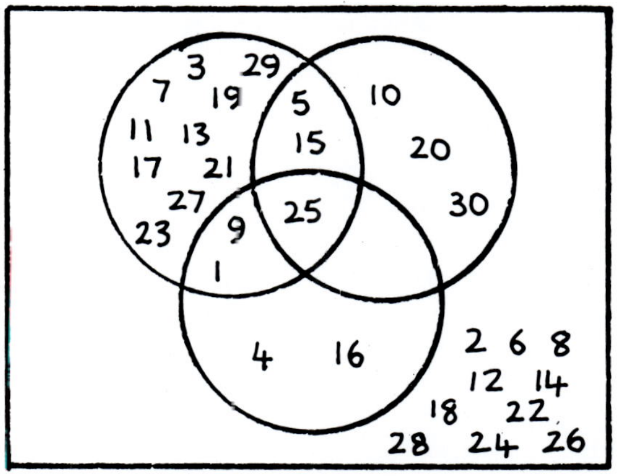

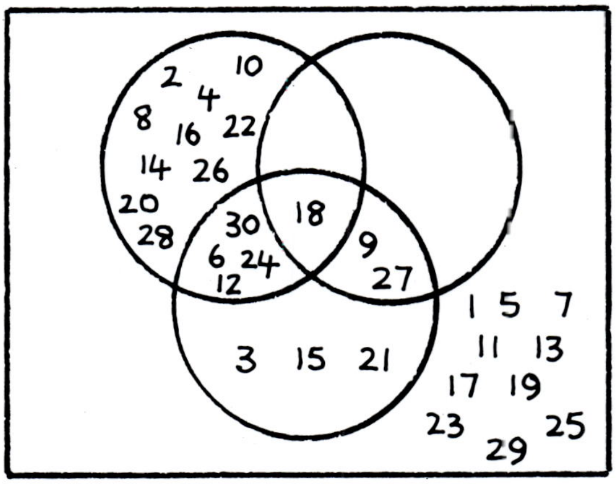

Page 90:
Repeated subtraction
9 children; 5 coins, 20 coins; 6

Page 91:
Equal sharing
8 strawberries; 25p;
16 in each set.

Page 92:
Halving
52; 23; 46; 35
59; 64; 29; 32

Page 93:
Doubling
25; 63; 85; 87

Subtracting
41; 41; 23; 31; 52

Page 94:
Remainders galore
The remainders are 0, 1, 2, 3, 4, 5, 6, 7, 8, 9

Zero remainders
1, 2, 3, 5, 6

Remainder Puzzle
Pear, grape, orange, banana

Page 95:
Remainder Problem
Any number greater than 4;
8, 13, 18, etc

Decimal Remainders
11.5; 6.4; 4.5; 8.25

Number Search
60, 90, 120, etc

Page 96:
Musical Division
Twinkle, Twinkle Little Star

Division Code
Pluto; Venus

Page 97:
Division Maze
There are 2 routes

Division Alphabet
Drum, Horn, Triangle (All musical instruments)

Page 98:
Equations
8; 8; 63; 9; 4; 42

Open Equations

1,24	3,1	1,36	5,1	1,21	8,1
2,12	6,2	2,18	10,2	3,7	16,2
3,8	9,3	3,12	15,3	7,3	24,3
4,6	12,4	4,9	20,4	21,1	32,4
6,4	15,5	6,6	25,5		40,5
8,3	etc	9,4	etc		etc
12,2		12,3			
24,1		18,2			
		36,1			

Missing Digits
4; 2; 8,4; 6,7

Final Digit
432, 747, 306, 846, 954

Page 99:
Function machines
Out: 3, 8, 9, 4, 7, 6
In: 63, 42, 49, 35, 56, 28

Table Grids (a possible answer)

×	9	7
4	36	28
6	54	42

×	5	8
9	45	72
10	50	80

×	4	8
3	12	24
7	28	56

Chains
64, 32, 16, 8, 4, 2, 1
4096, 1024, 256, 64, 16, 4, 1

Page 104:
Division Estimator
100; 100; 250; 350; 500

Page 105:
Calculator Quickies
2622; 12; r9

Two-digit Repeats
11; 66; 55; 33; 22; 99

Page 106:
Factor Trees

36	*48*	*32*
4 9	*6 8*	*4 8*
2 2 3 3	*3 2 2 4*	*2 2 4 2*
	3 2 2 2 2	*2 2 2 2 2*

Factor Query
12, 24, 36, 48

Page 107:
Divisible by 2
46, 124, 438, 522

Divisible by 5
95, 120, 435, 600

Divisible by 10
120, 600

Divisible by 4
236, 120, 332, 516
1544, 1840

Page 108:
Sieve of Eratosthenes
The prime numbers are:
1 2 3 5 7 11 13 17 19 23 29 31 37 41 43 47 53 59 61 67 71 73 79 83 89 97

Page 109:
Prime Quiz
True, true, true

Page 110:
Decimal Divisions
Possible dividing numbers include:
1 and 4, 1 and 8, 5 and 8,
3 and 4, 2 and 5.
.2, .4, .6, .8
25, .5, .75
.1, .2, .3, .4, .5, .6, .7, .8, .9
There are several possible answers to each problem.

Repeating Divisions
Dividing by 7, 9, 3 produces repeating decimals.
Dividing by 6 sometimes produces repeating decimals.

Page 111:
Dividing Fractions
0.375: 0.8: 0.5: 0 875: 0.3: 0.25

Nearly Fractions
4/7: 5/6: 7/9: 1/3

Page 113:
Best Buys
D1.50: D2.40: D1.95

Part Measures
115g. 10p.

Liquid Measures
80; 110; 150; 70

Page 114:
Carl; Marie; Simon and Tina

Index